Algorithms for Mobile Data Collection in Wireless Sensor Networks

RAJ ANWIT

Contents

List of Figures

Acronyms and Notations Used

Acronyms

ACO	Ant Colony Optimization
ANOVA	Analysis Of Variance
BS	Base Station
BFS	Breadth First Search
CC	Cluster Center
CI	Confidence Interval
CR	Consistency Ratio
CSS	Combine Skip Substitute
DTN	Delay Tolerant Networks
DA	Data Aggregation
ECR	Energy Consumption Rate
EDF	Earliest Deadline First
EV	Evolutionary
FS	Free Space
GA	Genetic Algorithm
HOAA	Heuristic Obstacle Avoiding Algorithm
HSD	Honest Significant Difference
IoT	Internet of things
ICP	Intersection Control Points
LEACH	Low Energy Adaptive Clustering Hierarchy
MEC	Mobile Edge Computing
MS	Mobile Sink
MP	Multi Path
MCPP	Mobile Collector Path Planning
MOPP	Modified Optimal Path Planning
MU	Manageable Units

MWSF Minimum Weighted Sum First

OP Odour Particles

OACE Obstacle Aware Connectivity Establishment

OF Objective Function

PRP Potential Rendezvous Points

PSO Particle Swarm Optimization

RP Rendezvous Point

SC Spectral Clustering

SCTR Spectral Clustering Tour

SAS Sweeping with Active Skipping

SN Sensor Nodes

SSO Shark Smell Optimization

STC Spanning Tree Covering

TSP Traveling Salesman Problem

TSPN Traveling Salesman Problem with Neighborhoods

WSN Wireless Sensor Network

Notations

n^s Number of SNs

n^r Number of RPs

C^p Set of ICP

n^h Number of points in C^p, i.e., $n^h = |C^p|$

r_0 Location of the BS

S Set of SNs, $S = \{s_1, s_2, \ldots, s_{n^s}\}$

R Final set of RPs, $R = \{r_1, r_2, \ldots, r_{n^r}\}$

c^s Communication range of SNs

$d_{i,j}$ Euclidean distance between r_i and r_j

$d^s_{i,j}$ Euclidean distance between s_i and s_j

$\aleph$ Set of potential RPs (PRPs)

D_{tour}	Path/tour of an MS
N_{tour}	Normalized tour length
$\lvert D_{tour}\rvert$	Path length or the tour length
D_{stour_i}	Subtour of an ith MS
n^{seg}	Number of network segments
n^{ns}	Average number of SNs in a network segment
n^{p}	Population size in Jaya algorithm
r^i_1	First RP of ith subtour
r^i_l	Last RP of ith subtour
Φ	Set of network segments, $\Phi = \{\delta_1, \delta_2, \ldots, \delta_{n^{seg}}\}$
S	Set of SNs, $S = \{s_1, s_2, \ldots, s_{n^s}\}$
R	Set of RPs, $R = \{r_1, r_2, \ldots, r_{n^r}\}$
w	Weight assigned to objective function components
α	Threshold value of strong signal strength reception of an SN
S^i	Set of SNs covered by r_i (in hop distance) in a path of a MS
$S^{i,j}$	Set of SNs covered by r_i (in hop distance) for the jth network segment
β	Maximum number of MSs, $\beta = \lceil \alpha_c \times n^r \rceil$
v	Optimal number of MSs, $v < \beta$
Λ	Number of MSs, $v \leq \Lambda \leq \beta$
c^s	Communication range of SNs
$n^h_{i,j}$	Number of SNs in i^{th} convex hull of jth network segment
n^{th}	Threshold value for number of RPs in a partition
n^t_i	Number of RPs in D_{stour_i}
s^r	Reference SN of a network segment
n_{r_i}	Number of SNs covered by the RP r_i
mu_i	i^{th} manageable unit
D_{mu_i}	Degree matrix of mu_i
x^I_i	Indicator vector of size $\lvert mu_i\rvert$
n^{st}	Number of sets
C	Set of subsets of S, $C = \{\phi_1, \phi_2, \ldots, \phi_{n^{st}}\}$
R_{stour_i}	Set of RPs in i^{th} subtour

Symbol	Description
$\mathscr{F}$	Set cover of C
α_c	A constant to estimate maximum number of MSs
c_i	Cost associated with i^{th} set
x_i	Boolean variable for choosing sets
k_{max}	Maximum number of iterations
A_{mu_i}	Adjacency matrix of mu_i
k	Number of partitions of mu_i
E_i	Set of edges of mu_i
V_i	Set of vertices of mu_i
U	SNs of mu_i in k dimensional vector space
Z_i	i^{th} row of U
p_{val_i}	post order traversal value of S_i
ψ	Set of connected components
Γ	Standard deviation of subtour lengths
δ_i	i^{th} network segment
S_i^s	Set of SNs in i^{th} network segment
h_i	i^{th} convex hull
S_i^h	Set of SNs included in a convex hull h_i
μ	Mean of the lengths of all subtours
v_m	Average speed of a MS
D	Data collection latency

Chapter 1

Introduction

Chapter 1

Introduction

1.1 Background

The last few decades have witnessed ubiquitous spread of wireless sensor networks (WSNs) to a vast range of data-intensive applications, ranging from precision-agriculture [1], health-care [2], forest fire detection [3], surveillance system [4], military [5], and so on. Typically, a WSN is made up of spatially distributed autonomous devices called sensor nodes (SNs) that communicate wirelessly, collecting data, and detecting significant events in physical and environmental circumstances [6]. Each of these SNs has the ability to concurrently sense, process, and communicate data to a base station (sink). SNs route the data to the sink via single-hop or multi-hop communication to the base station (BS) or the sink [7]. However, SNs are resource constrained in terms of power supply due to limited battery capacity [8]. Therefore, energy efficiency and thus network lifetime are the critical research objectives due to battery limitations. With large scale WSNs a reality in a variety of real-life applications, the uneven energy dissipation is the "killer" of network lifetime. One such situation is as follows. In a traditional WSN with a static sink, the traffic of delivering data to the BS from SNs in the entire network follows a many to one (converge-cast) traffic pattern [9] as shown in the figure 1.1. As a result, the SNs neighboring the sink node face congestion, as they are frequently used for delivering the data to the BS. Consequently, these nodes deplete their energy quicker than other far away SNs leading to the early death of the network [6]. This situation is often referred by the literature as the energy sink-hole problem [9] or the energy-hole/hot-spot problem [10–12]. Extensive research has been carried out to address this issue, e.g., efficient communication protocols [13–15], multiple sink systems [16], optimized network deployment [17, 18], and so on. However, the problem cannot be fully

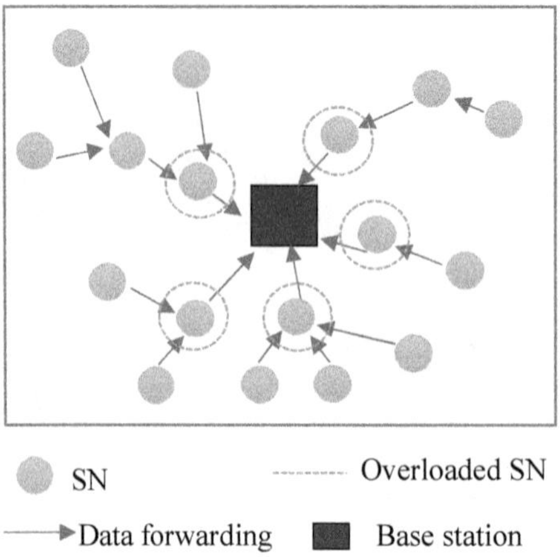

Figure 1.1: An example demonstrating the sink hole problem.

tackled as long as the SNs and the sink are static. To this end, exploiting sink mobility has emerged as a potential solution to enhance multiple performances in WSNs such as network lifetime, energy consumption, and data collection latency [9, 10, 19–21]. In other words, sink mobility provides researchers an edge in the design of schemes that could potentially lead to better solutions. In the meantime, the overhead of network dynamics due to mobility can degrade the performance of the network [19]. Therefore, it must be managed in such a way that the associated benefits outweigh the damages caused.

Mobile Sinks (MSs) are resource-capable entities predominantly used for data collection from the nearby SNs in one-hop or limited multi-hop communication inside a WSN. The capability of resources refers to the sufficiently large reserves of energy and memory for gathering data. An MS periodically visits some sojourn geographical locations in the network called rendezvous points (RPs) [22] for data collection. At these locations, the SNs, which are mostly one-hop away, send their data to the MS upon its arrival. As a result, long-hop relaying is circumvented, the total energy consumption of the network is reduced, and the network lifetime is enhanced. Fig. 1.2 shows a working example of an RP-based data collection scheme using MS. Note that several issues are to be addressed when employing MS(s) depending on the application and nature of the WSN as follows.

(a) How to choose the number and position of the RPs? The nature of this selection problem such that the trajectory of the MS becomes optimal is considered to be NP-hard in nature [23]. The task becomes even more challenging in a large class of

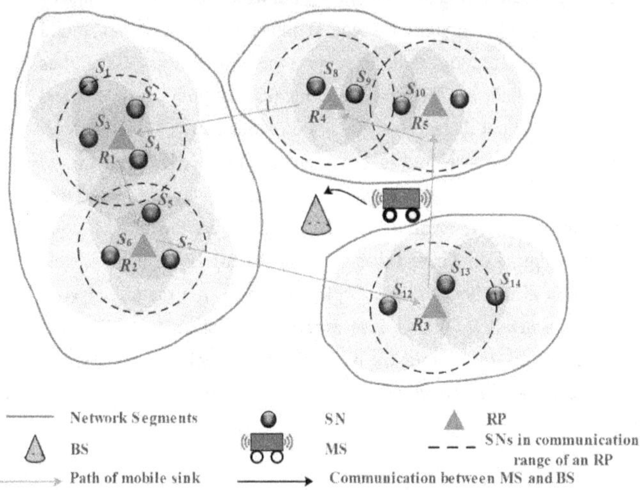

Figure 1.2: A working example of an RP-based data collection scheme using MS.

WSNs applications such as agriculture, wildlife monitoring, and perimeter security which are normally populated with obstacles of all sorts, including small water bodies, large stones, tall trees, forest areas, damaged land, or infrastructure, etc.

(b) In a WSN, a single MS might be sufficient for data collection depending on the application. For example, applications such as forest wildlife monitoring require data once every few weeks (delay-lax). Meanwhile, single MS might be insufficient for data collection in applications such as military perimeter monitoring, which requires data periodically after a few minutes (delay-harsh). It is noteworthy that delay tolerant networks (DTNs) usually have sparsely distributed SNs with no persistent connectivity guarantee. Therefore, the next question arises: how to find the minimum number of MSs and their associated tours such that data can be collected from all the SNs with minimum tour length and latency.

Both of the above questions, when tackled separately or when considered sequentially, have to consider a trade-off between network delay and energy efficiency as they consequently affect the network lifetime. Additional consideration that needs attention is when the WSN has disjoint portions which has become a reality in existing application.

Nowadays, WSNs are also the most important component of the Internet of Things (IoT) [24] that perform data collection, which is the main task of IoTs applications [25]. Note that all IoT applications overlap with WSN applications as WSNs are the underlying infrastructure in IoT. They generate a gigantic amount of data which crucially becomes a burden to manage and analyze [26]. To cater the needs of this vast amount of data, cloud computing is the most emerging technology that can support the limited storage, analytics demands, and large-scale computations. Unfortunately, it causes significant delays and transfer costs due to the far physical distance of the cloud data center [27]. Therefore, mobile edge computing (MEC) is considered in IoT scenarios [28]. MEC can extend cloud computing services to the network edge so that data transfer delay is reduced and response time is much shorter [29]. An example of a three-layered architecture of cloud edge-based IoT is shown in Fig. 1.3. There exists cloud at the top layer, edge servers at the middle layer, and IoTs at the bottom layer, which is shown in a magnified view immediately right to it. In a nutshell, WSNs have evolved to exist as data collection entities in this architecture. In this architecture, the data collection problems posed earlier still exists and demands considerable effort to design. The task becomes even more difficult when obstacles are present in the environment.

1.2 Research Gap

Extensive research efforts are made to develop RP based data collection schemes which can be classified as single-hop based data collection [12,30–34] and multi-hop based data collection [21,22,35–52]. In the former scheme, only single-hop communication between SNs and MS(s) are permitted, whereas in the latter scheme multi-hop communication is allowed. On the one hand, single-hop schemes allow the benefits mentioned previously, while the multi-hop schemes may still lead to energy overhead for some SNs. Nevertheless, there are some schemes for single-hop based data collection schemes [53–55] wherein the MS visits each and every SN and gathers its data in single-hop communication. Although this limits the energy consumption of the nodes, the tour length of the MS becomes unreasonably long, which in turn results in high MS energy expenditure and high data collection delay causing nodes' buffer overflow. RP-based data collection schemes enable MS to collect a large volume of data by achieving a desirable balance between network energy saving and data collection delay. However, finding the optimal number of RPs and their locations for designing an efficient trajectory for the MS is challenging. To this end, various studies [30, 35, 38] have adopted clustering as an effective approach. However, these techniques have several shortcomings. Ghosh et. al. [30] have proposed a method for finding the number of sojourn positions (i.e., RPs) so that a set of mobile collectors (i.e., MSs) traverse these RPs for data collection from SNs. However, their method has not considered finding the optimal number of RPs, which results in suboptimal tours for the set of MSs. The proposed method in [35] has also applied clustering called MobiCluster to determine the number and locations of the RPs. Nevertheless, there is still scope for optimizing the number of RPs as well as overall energy consumption. Similarly, the proposed method in [38] produces clusters with uneven distribution of SNs, thereby causing unbalanced energy consumption of MS at RPs.

Next, the schemes mentioned above have not considered the simultaneous optimization of RPs and MSs. This may not be overlooked because using an optimal number of RPs and MSs can not only reduce the data collection delay significantly but also reduce the energy consumption of the SNs, especially in large-scale networks. However, it is challenging to determine an optimal number of RPs and so the optimal path for each of the MSs. The task becomes more demanding in delay tolerant networks (DTNs), which usually have sparsely distributed SNs with no persistent connectivity guarantee. Although DTNs are not time-critical, they are characterized by diverse delay requirements. For example, forest wildlife monitoring applications require data once every few weeks (delay-lax DTN); on the other hand, military perimeter monitoring requires data periodically after a few minutes (delay-harsh DTN) [40]. Hence, there should be a trade-off between network delay and energy efficiency while finding out an optimal number of the RPs. The literature [32, 36] has proven this problem as NP-hard since there are $\binom{n}{m}$ possible selection of m RPs from n SNs. For the large-scale WSNs, the computational complexity of this problem would be non-polynomial. To address such problems, nature-inspired computing paradigm based meta-heuristic algorithms are proven to be more promising in the literature for developing efficient solutions [33,34,44]. Specifically, [33] and [44] are based on the ant colony optimization (ACO) algorithm, wherein they have considered a multi-hop communication scenario between MS and SNs. [34] is based on the particle swarm optimization (PSO) algorithm, which aims at optimizing the tour for a single MS only. This scheme lacks scalability, which prolongs the data collection delay in large-scale networks. Also, they have not considered jointly optimizing the number of RPs and MSs.

Still further, the propagation model in the existing schemes is assumed to be oversimplified as they do not follow the signal path-loss model. This should not be overlooked as the signal strength drops considerably when an entity moves away from its center, hampering the real-time data collection. Therefore, a realistic propagation model that imbibes the log-normal shadowing phenomena is overriding to design data collection schemes for real-life applications [56]. Additionally, the works [9,22,31,32,36,38,40,46–48,57,58] have to be re-analyzed, as none of them has considered path planning of the MS in the presence of obstacles. This may be noted that a large class of IoT applications such as agriculture, wildlife monitoring, and perimeter security is normally populated with

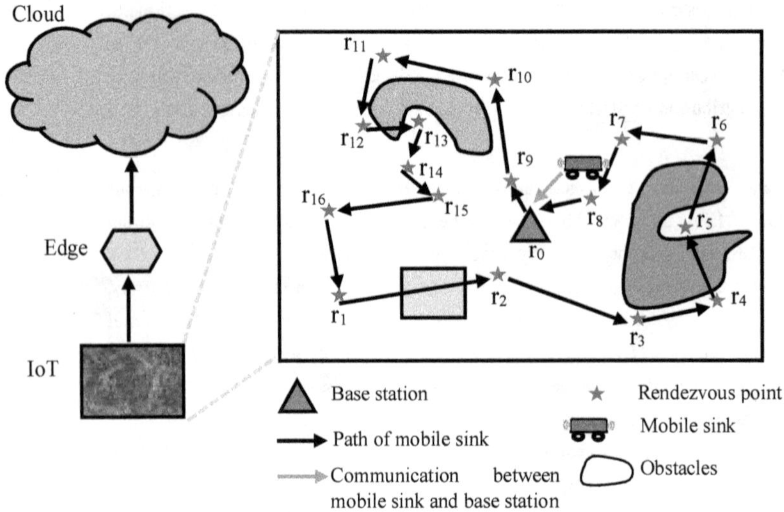

Figure 1.3: An example of three layer architecture involving edge. The IoTs are essentially WSNs deployed at the edge of the network.

obstacles of all sorts, including small water bodies, large stones, tall trees, forest areas, damaged land or infrastructure, etc. An example scenario is shown in Fig. 1.3, which has three different obstacles indicated using three different colors. The presence of obstacles makes the topological organization of the network complex, and communication between different network entities is also obstructed. Herein, several issues are to be addressed. 1) The shape of the obstacle is highly irregular, so some SNs may fall in the convex part of the obstacles. In this case, it is difficult to determine RPs which lies inside the convex part (i.e., r_5 and r_{13} in Fig. 1.3). 2) Once the RPs are determined, the path of the MS is planned (i.e., $r_0 \rightarrow r_9, \ldots, r_8 \rightarrow r_0$). Unfortunately, this path may pass through the obstacles. For example, the path is obstructed four times in Fig. 1.3, i.e., at $\overline{r_{12}r_{13}}$, $\overline{r_1r_2}$, $\overline{r_4r_5}$, and $\overline{r_5r_6}$. As a result, a partially feasible path is obtained that cannot accomplish data collection. 3) Next, the constructed path may also have sharp turns at some RPs and at the intersection of the path with the obstacles. For example in Fig. 1.3, the four intersection points of the path segments with the obstacles and r_{15}, r_{16}, r_1, etc. bill sharp turns. The consequence is that the MS faces jerks which affects the quality of movement as well as the hardware of the MS. Moreover, such paths are associated with high energy consumption and high latency. Therefore, optimal path planning of the MS for WSNs with obstacles is very challenging in several IoT applications, as stated above.

In general, single-hop communication with an optimal number of RPs is paramount to conserve the energy of SNs and enhance the network lifetime for sustainable operation of applications. However, in delay-sensitive applications, this approach is not desirable. Thus, single-hop communication coupled with an optimal number of MSs is like 'the icing on the cake', which annexes the following advantages to the applications:

- Timely delivery of data, supporting their time-boundedness/ time-sensitiveness nature.
- Shielding from loss of data due to buffer overflow of SNs.
- Detection of node failures in regions that are sensitive to monitoring.
- Total energy consumption reduction and network lifetime enhancement.

Now, the above features, when joined with obstacle avoidance and smooth path construction, endows the WSNs with the following additional benefits along with the advantages mentioned above:

- Uniform speed support for the MS(s), making them meet their goal in time.
- Longer operational time for the MS(s) by preventing hardware damage and conserving energy.

It is to be noted that finding an optimal number of RPs in the presence of highly irregular-shaped obstacles is very challenging. However, once some mechanism is established for this, the underlying network can enjoy all the benefits discussed above.

1.3 Research Contributions

The thesis mainly presents data collection schemes based on single-hop RP-based communication by MS(s). Here, we present a brief description of each of the research contributions in the thesis as follows.

1.3.1 Heuristic Based Data Collection Scheme

In the first contribution, we present a novel scheme for trajectory design of mobile sink for data collection. The problem considered here is to optimize the number of RPs and their locations with a goal to minimize the traveling length of the MS. First, we reduce the problem size using a combination of breadth-first search and Tarjan's algorithm [59] and then apply spectral clustering (SC) [60, 61] to find the optimal set of rendezvous points to plan the tour for the mobile sink. The scheme is based on SC and thus called spectral clustering tour (SCTR). We have performed extensive simulations, and the results are compared with relevant existing schemes. The comparative results confirm the effectiveness of our approach in terms of the number of rendezvous points, path length, the variance of rendezvous points, and energy consumption per round. The validation of simulation results have also been shown through statistical t-test [62].

1.3.2 Shark Smell Optimzation based Data Collection Scheme

In the second contribution of the thesis, we present a novel scheme comprising of two algorithms based on the shark smell optimization (SSO) [63] technique that solves the MS tour planning problem. The problem considered is to find the optimal number of MSs such that the associated tour lengths with MSs are balanced. To solve the problem, the first SSO-based algorithm is used to determine an optimal number of RPs and their locations. Based on this, the second algorithm optimizes the number of MSs so as to minimize the overall tour length. Each of the algorithms is developed with an efficient and novel particle encoding scheme along with the derivation of the fitness function. The tour planning is formulated as an Integer Linear Programming problem for the first algorithm and a Non-linear Programming problem for the second algorithm. Simulation results of our scheme confirm the improvement over the state-of-the-art algorithms. The results are also statistically validated through hypothesis testing using ANOVA [64] and post hoc analysis [62].

1.3.3 Convex Hull and Jaya Algorithm Based Data Collection Scheme

In the third contribution of the thesis, we introduce WSNs as the central means to disseminate data for Internet of Things (IoT) applications in the era of mobile edge computing. Herein, we consider a disjoint network scenario found in many IoT applications. We also consider a realistic propagation model for communication between different network entities. The issue at hand here is same as that discussed in Section 1.3.2, i.e., optimizing the number MSs such that the associated tour lengths with MSs are balanced. First, we propose a convex hull-based algorithm to determine RPs for constructing an optimal tour of an MS. Then, using the resulting set of RPs, we present another algorithm based on the Jaya metaheuristic [65] to determine an optimal number of MSs and their balanced tours. We name our proposed scheme Convex Hull and Jaya Algorithm-based Data Collection (CHJADC). Rigorous simulations show that our scheme outperforms existing algorithms in terms of various performance metrics. We also statistically validated the simulation results using ANOVA and post hoc analysis.

1.3.4 Heuristic Based Obstacle Avoiding Data Collection Scheme

In the final contribution, we again present WSNs as the predominant channel for data dissemination in edge-based IoT. Herein, the MSs may face obstacles, which become hurdles in data collection. Most of the existing works have addressed the problem of path planning for MS in the presence of obstacles but without any concern for constructing an optimal path. Moreover, none of them has considered the following issues dealing with obstacles, a) optimizing the number of RPs and constructing a viable path and b) smooth path construction by considering sharp edges and convexity of the obstacle topology. Therefore, the problem considered here to find optimal number of RPs in obstacle ridden network with the aim to reduce the path length, total energy consumption, and data collection latency. After that, smooth path construction is tackled. In this chapter, we address the above mentioned issues and develop an efficient scheme for the determination of an optimal number of RPs by a greedy approach solution to the set-cover problem and construction of the optimized path in polynomial time. Then we apply the modified BUG2 [66] algorithm to construct an obstacle-free path, which is then made smooth using the concept of the Bezier curve. The modified path makes the data collection scheme efficient in terms of energy and latency. The scheme is called Obstacle Aware Smooth Path Planning (OAsPP). Extensive simulations show the superior performance of the proposed scheme over existing competing algorithms in terms of various performance metrics.

1.4 Organization of the Thesis

The organization of this thesis is outlined in Figure 1.4. Chapter 2 presents the detailed review of relevant and competing researches on the data collection using MS(s) in WSNs. Chapter 3 discusses a heuristic-based data collection scheme, including the problem statement and simulation results. Next, the Chapter 4 presents the SSO-based data collection scheme. Chapter 5 presents the details about the convex hull and Jaya algorithm-based data collection scheme via edge computing. After that, Chapter 6 presents the heuristic-based obstacle avoiding data collection scheme in the Internet of Things. Finally, Chapter 7 wraps up the thesis by putting forward the future research directions and limitations of our work.

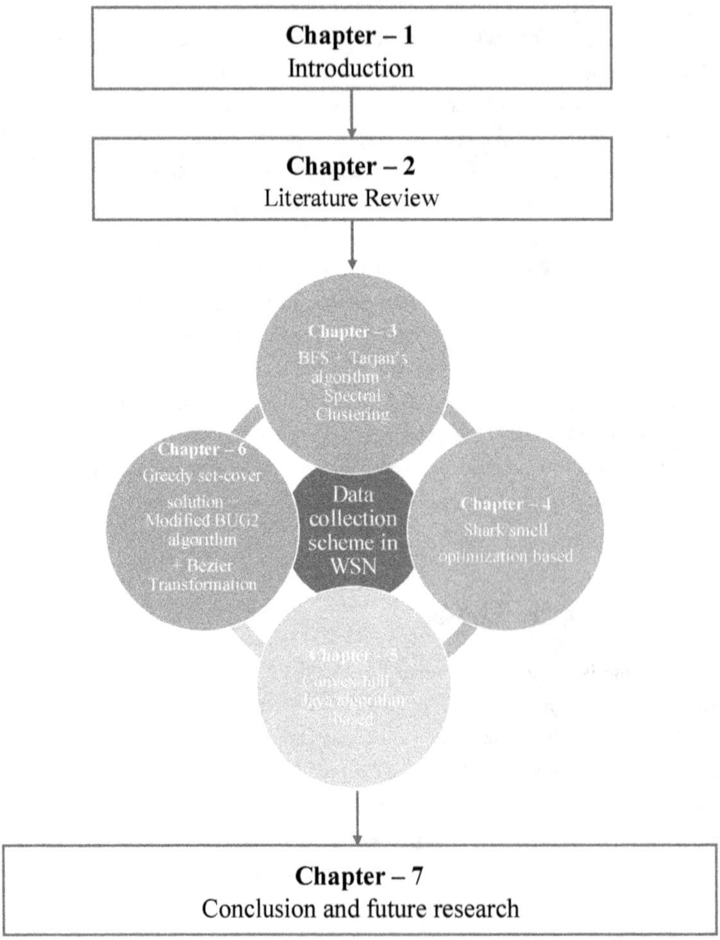

Figure 1.4: Organization of the thesis.

Chapter 2

Literature Review

Chapter 2

Literature Review

2.1 Introduction

There is rich literature available for the data collection approaches in WSNs, which are relevant and competing with our work. The MS may visit each and every SN and collect their data in single-hop communication [53–55] or may visit only some geographical locations of the WSN, and the SNs send their data to the MS through single-hop or multi-hop communication [12, 21, 22, 30–52]. In the former solution only single-hop communication is required, however, the total energy consumption is low and data delivery delay is high. In the latter approach, the characteristics are complex and depend on the application involved. However, some generalizations can be made as 1) The data collection latency delay is low. 2) Total energy consumption due to limited multi-hop communication is relatively high. Overall, a high network lifetime can be achieved by considering the trade-off balance between delay and energy.

The schemes can be grouped into two categories based on the algorithms involved: heuristic-based and hybrid approach-based. In the first classification, some heuristic-based approach is involved for the design of the data collection [12, 21, 22, 30–32, 35–43, 45–50, 52, 67–69]. In other words, these works are based purely on heuristic methods. Such approaches provide efficient and quick solutions. However, such solutions portray better performance with respect to application-specific scenarios. The well-known metaheuristic approaches such as particle swarm optimization [70, 71], ant colony optimization [72], genetic algorithm [73], etc., have been utilized to design a data collection scheme [21, 33, 34, 44] in the second category. These approaches have been proven by literature to provide near-optimal solutions to NP-hard problems. However, such solutions may also deviate from the optimal solution space. Thus, the metaheuristic algorithm is combined with a suitable heuristic approach to amalgamate the benefits of the above two approaches to obtain better solutions for data collection. In short, these works are a combination of heuristic, metaheuristic, or some other methods, utilizing the benefits of each and aiming to achieve good enough solutions. A comparative study of all the above schemes has been presented in the Table 2.1.

2.2 Heuristic Based Data Collection Schemes

Herein, the heuristic-based data collection schemes have been discussed that are most relevant to the work in this thesis. A brief description of each of the schemes have been provided so that their shortcomings are highlighted and research gap is evident.

In [32], authors have proposed a method that reduces the length of the traveling path of the mobile sink. Nevertheless, the computational overhead is high. Additionally, the scheme is not suitable for large and delay-sensitive applications. In [36], authors have proposed a spanning tree covering algorithm which uses M-collector(s) to plan a tour for the MS with the help of polling points. Herein, a spanning tree is constructed based on a greedy heuristic method in each of the network partitions. This spanning tree is then decomposed into subtrees, and approximate tours are formed for each of these subtrees. Different tours are assigned to M-collector for collecting data. In [39], a routing tree is constructed, and afterward, each edge is assigned a weight based on the numbers of SNs to relay the data to the sink. The MS is restricted to move along the edges, implying that an SN is visited twice, which makes the path of the MS inefficient. Herein, multi-hop communication is allowed, and SNs serves as the RPs. In [53], two algorithms are presented: earliest deadline first (EDF) and minimum weighted sum first (MWSF). Both of these algorithms require the MS to visit each of the SNs of the network, which makes it practically infeasible for large WSNs. Moreover, the energy requirement and data collection latency of the scheme is relatively high. In [54], an MS is used to collect data from all SNs. The problem is termed the data mule scheduling problem, where the data mule has to visit each and every SNs by determining the appropriate speed and the sequence of visits. Then, the shortest path is formed to reduce data delivery latency.In [55], there are two proposed algorithms: TSP with neighborhoods (TSPN) and evolutionary-based TSPN (EV-based TSPN). These algorithms too consider that each and every SN must be visited, eventually making the path length of the tour unacceptable for large networks. In summary, these schemes either increase the data collection delay or consume a large amount of energy to transmit data to the MS. These schemes fail to sustain the longer operation of the WSNs.

The authors in [68] have presented an algorithm called sweeping with active skipping (SAS). The scheme at first considers all SNs as RPs, then combines the SN whose communication range overlaps. This scheme is a modification of the Welzel algorithm [74]. However, the number of visiting points is not optimal, leading to plenty of scopes for their optimization and improving the path length.

In [69], the proposed algorithm uses a cluster-based method to find RPs. The scheme constructs balanced-size clusters such that SNs in each cluster are nearly the same. After that, the cluster heads act as RPs, which are used to plan a tour for the MS. However, the tour failed to pass through the parts of the network having higher node density. Hence, the data forwarding follows longer paths leading to non-uniform energy consumption and reduced network lifetime of the WSN. In [35], an uneven cluster-based algorithm is proposed where clusters near the BS have a smaller size and the size increases with distance from the BS. The faraway cluster heads face high data relaying load, and the path is inefficient. Moreover, the faraway clusters may have multi-hop communication, which spawns uneven energy consumption. Also, this scheme may not be suitable for large-scale WSNs. In [38], clusters are formed based on the grid size, and the center of the grid serves as an RP. Here, the optimal grid size is difficult to estimate, and there is an unbalanced number of SNs in case of random deployment of SNs. Note that the optimal grid size determination is an overhead. Hence, the optimal number of RPs cannot be determined correctly. This leads to the uneven energy consumption of SNs and reduced network lifetime. Also, there is no concern for finding out the optimal number of MSs. In the proposed method in [30], clusters are formed by diving the network with the help of an equilateral triangle, and the cluster centers are designated as RPs. In this case, the number of clusters is not optimal, and hence the optimal set of RPs cannot be found, leading to a suboptimal path of the MS. In [75], a clustering-based data collection method has been proposed where there are multiple numbers of cluster heads in each cluster. In this algorithm, the number of clusters is not optimal, and the number of SNs per cluster is not balanced, which leads to an inefficient path for the MS. In [76], the proposed algorithm allows multi-hop communication in clusters, so the SNs near the cluster center spend relatively high energy. Another problem is that finding optimal sojourning locations of the MS is not considered. In [40], the authors present a quick convex hull-based data collection scheme for delay harsh disjoint WSNs. In the scheme, there are several disjoint network segments. From each network segment a representative location is selected as an RP. Those RPs are then used to plan a path for the MS using the proposed convex hull approach. However, the number of RPs in the scheme is not optimal and has a lot of scope for improvement.

The authors in [77] proposed a path planning scheme where clustering is performed first, and then Dijkstra's shortest path algorithm is employed. In [78], the authors employed a Delaunay triangulation-based obstacle avoidance clustering method to plan a path for an MS. In [79], the authors proposed a cut-vertex-based approach to plan a path for mobile robots. The authors in [80] proposed a scheme of path planning involving multiple robots. In [81], the authors proposed a path-planning scheme using a hybrid search based on $E*$ and Bernstein-Bezier motion primitives, which enables a faster motion of the robot. The resultant path obtained using the scheme is shorter, smoother, easy to traverse, and has lower computational complexity. The authors in [82] proposed a scheme for path planning by avoiding obstacles based on Hamilton-Jacobi-Bellman Equation. In all of the above schemes, only regular-shaped obstacles are considered, and smooth path construction is missing. These types of obstacles are relatively easy to handle and do not encompass real-life applications where obstacles are very irregular in shape.

In [83], the authors at first proposed a clustering-based algorithm to find RPs, then heuristic based on spanning tree for tour-planning using a single MS to find the obstacle-avoiding shortest route. The algorithm is called a heuristic obstacle avoiding algorithm (HOAA). However, it suffers from sharp turns and fails when an RP falls inside the convex part of the obstacle. In [84], the authors proposed B-spline curves for the path planning problem. However, the inherent property of B-spline curves is that they have the local best performance. Thus, there is significant scope for their improvement. In [85], the authors proposed a scheme based on fast distance transformation and variations of some topological methods as thinning and skeletonization to plan a path for mobile robots. In [86], the authors proposed a scheme that is based on a path graph optimization technique that employs a compact mesh representation. Then, a vertex pushing method is applied iteratively to include important obstacle boundary edges in the path graph. Finally, Dijkstra's shortest path searching algorithm is used to find the shortest path. The authors in [87] proposed a scheme based on the construction of a graph where edges that are assigned weights based on the remaining energy of each SN. The graph is then used to find the shortest paths. In [88], the authors presented an efficient heuristic-based path planning scheme in the presence of obstacles. In [89], and [90] the authors have proposed similar schemes for planning a path for a robot or an MS or a vehicle by avoiding obstacle. In [91], the authors proposed a scheme that comprises gliding path generation, improved A star algorithm, and segmented Bezier curve smoothing for path planning for underwater applications. All of these above-mentioned schemes use obstacles of arbitrary shapes. A summary of the selected heuristic data collection schemes evaluated on various parameters has been shown in Table 2.1.

Table 2.1: Summary of the heuristic based data collection schemes

Reference	Direct or RP-based	Optimization of RPs and MSs	Obstacle avoidance	Realistic propagation model	Network type	Heuristic used
[54]	Direct	NA	No	No	Connected	Based on time interval
[53]	Direct	NA	No	No	Connected	Graph-based scheduling
[55]	Direct	NA	No	No	Connected	TSP optimization based
[32]	RP-based	No	No	No	Connected + Disconnected	TSP and COM[74] algorithm based
[39]	RP-based	No	No	No	Connected	Graph traversal based
[36]	RP-based	Yes	No	No	Connected + Disconnected	Spanning tree based
[68]	RP-based	No	No	No	Connected + Disconnected	Sweeping + COM[74] algorithm based
[69]	RP-based	No	No	No	Connected + Disconnected	Clustering and binary search
[30]	RP-based	No	Yes	No	Connected + Disconnected	Clustering and equilateral triangle based
[35]	RP-based	No	No	No	Connected + Disconnected	Uneven clustering based
[38]	RP-based	No	No	No	Connected	Grid cell division based
[75]	RP-based	No	No	No	Connected	Clustering-based
[76]	RP-based	No	No	No	Connected	Clustering-based
[77]	RP-based	No	Yes	No	Connected	Clustering+ Dijkstra shortest path algorithm based
[78]	RP-based	No	Yes	No	Connected	Delaunay triangulation based
[79]	NA	No	Yes	No	Connected	Gyroscopic force and cut-vertices
[80]	NA	NA	Yes	NA	Connected + Disconnected	Grid map and region searching
[81]	NA	NA	Yes	NA	Connected + Disconnected	Hybrid E^* algorithm and Hybrid E^* Bernstein-Bezier motion primitives
[82]	NA	NA	Yes	NA	Connected + Disconnected	Hamilton-Jacobi-Bellman (HJB) theory
[84]	NA	NA	Yes	NA	Connected + Disconnected	Quartic uniform B-spline curves
[85]	NA	NA	Yes	NA	Connected + Disconnected	Fast distance transformation (FDT) and thinning and skeletonization
[86]	NA	NA	Yes	NA	Connected + Disconnected	Path graph optimization using compact mesh representation
[87]	Direct	No	Yes	NA	Connected + Disconnected	Energy-based Voronoi diagram and digraph
[88]	Direct	No	Yes	NA	Connected + Disconnected	Virtual-force-based method and floor- based scheme
[89]	RP-based	No	Yes	No	Connected + Disconnected	Clustering-based
[90]	Direct	NA	Yes	No	Connected + Disconnected	Bipartite graph
[91]	Direct	NA	Yes	No	Connected + Disconnected	Gliding path generation+ Improved Astar (A^*) + segmented Bezier curve smoothing.

2.3 Hybrid Approach Based Data Collection Schemes

In this section, we first discuss the works based on purely metaheuristic algorithms; then, we discuss the works based on a combination of heuristic and metaheuristic algorithms.

Very few works [34] exist that involve only a metaheuristic approach to perform data collection using MS in WSN. In [34], a variable dimension-PSO algorithm is proposed for finding out the RPs to plan a tour for the MS. Herein, a variable dimension particle is used to express one of the feasible paths, and the position of every RP on the path becomes one single dimension of the particle. The position of RPs is merged in those cases wherein the distance between two RPs falls below an assumed threshold value. In this way, an optimal number of RPs and their positions are achieved. However, the scheme leads to a large number of RPs and is not suitable for large-scale WSNs.

The following schemes involve a combination of heuristic and metaheuristic approaches. In [92], the authors proposed an improved ant colony optimization-artificial potential field hybrid algorithm with adaptive early warning for path planning of unmanned aerial vehicles. In [44] the authors proposed a mobile sink path planning scheme using the ant colony optimization (ACO) algorithm [93] for data collection constrained by variable data rates. The scheme first constructs a directed spanning tree using a weight assigned to the SNs. Then an optimal number of RPs are selected using an ACO-based algorithm. These RPs are finally used to plan a tour for the MS.

However, the scheme allows multi-hop communication between SNs and MSs, which is not desirable for the sustainable behavior of a WSN. The scheme may not be suitable for large-scale WSNs as a single MS is involved in data collection. The authors in [33], the authors first perform clustering of the SNs using a modified-LEACH algorithm. The cluster heads are designated as RPs, which are then used to plan a path for multiple MSs using an ACO-based approach. However, the approach suffers from the following limitations. 1) The clustering is not optional, resulting in a sub-optimal number of RPs 2) The resulting tours of the MS have a lot of scope for improvement. In [94], the authors proposed an energy-efficient multi-objective PSO-based path planning scheme for data collection using MS. At first, the RPs are obtained based on the overlapping communication range of the SNs. Then, those RPs are used to obtain a TSP tour for the MS. However, the scheme involves a single MS and thus may result in longer data collection delay making it unsuitable for large-scale WSNs involving time-critical applications. In [95], a PSO-based data collection scheme using MS is proposed, which allows multi-hop communication making it unsuitable for large-scale WSNs. Quite a few metaheuristic-based data collection approaches exist [67, 96–98] which involve MS, yet the most relevant ones are discussed above. These schemes also have not considered the optimization of the number of RPs and MSs in the model involving one-hop communication between SNs and MSs. Moreover, a realistic communication model is missing, which is crucial in real-life applications.

In [99], the authors proposed a tour planning scheme for mobile data collection in partitioned WSNs, which involves a combination of a heuristic and genetic algorithm. A potential RP from each network segment is selected using a heuristic measure; then, a reference center point is selected inside the network. After that, using a local adjustment of the RPs, the final position of RPs is chosen, and this is used to plan a tour for the MS. This tour is then optimized globally to reduce the length using the genetic algorithm. However, the following limitations are notable. 1) Multi-hop communication is possible in each network segment. Thus, the possibility of an energy hole problem still exists. 2) The number of RPs is not optimized. 3) The scheme will not work in time-critical applications. In [21], the authors proposed a path planning scheme that is designed especially for disconnected WSNs using a heuristic and ACO-based algorithm. The scheme is divided into three phases: In the first phase, a single potential RP from each grid is selected, and then an initial path or the path skeleton is developed. In the second phase (candidate grouping mechanism), the segments of paths obtained in the first phase are grouped to form a complete path. The last phase, called ACO-Based Path Formation, optimizes the complete path so as to minimize the tour for data collection. Similar approaches can be found in [43] where a heuristic approach is employed to find out the

RPs and then plan a path for the MS.

Table 2.2: Summary of the hybrid based data collection schemes

Reference	Direct or RP-based	Optimization of RPs and MSs	Obstacle avoidance	Realistic propagation model	Network type	Method used
[21]	RP-based	No	No	No	Connected + Disconnected	Convex hull and ACO based
[33]	RP-based	No	No	No	Connected	Clustering and Ant colony optimization based
[34]	RP-based	No	No	No	Connected	Variable dimension PSO-based
[41]	RP-based	No	No	No	Connected + Disconnected	ACO and tree based
[44]	RP-based	No	No	No	Connected	Ant colony optimization and spanning tree based
[67]	RP-based	No	No	No	Connected	Clustering and PSO based
[92]	RP-based	No	No	No	Connected	Ant colony optimization-artificial potential field (ACO-APF) algorithm + Grid map
[94]	RP-based	No	No	No	Connected	Multi-objective PSO + COM algorithm
[95]	RP-based	No	No	No	Connected	Artificial bee colony and particle swarm optimization
[96]	RP-based	No	No	No	Connected	PSO-based
[97]	RP-based	No	No	No	Connected	PSO and compression sensing theory
[98]	RP-based	No	No	No	Connected	GA based
[99]	RP-based	No	No	No	Connected + Disconnected	Convex hull and GA based

A common characteristic of all the aforementioned schemes in both the classifications is that initially, RPs are established with the help of some heuristic, cluster structure, or metaheuristic technique, and afterward, the tour of the MS is planned. The delivery of data between SNs and MS is in one-hop communication; however, some of the schemes allow multi-hop communication also. Most of these schemes are centralized approaches that try to minimize an energy, network lifetime, or path length-related cost function without paying proper attention to the selection of RPs. Specifically, they don't consider the number and location of RPs; instead, the focus is on RP selection to provide coverage of SNs. Also, there is no particular focus on the number of MSs. As a result, the path length becomes sub-optimal, and the delay in data collection becomes high. The main trade-off should be between the path length and delay in data delivery. Note also that many of the previous schemes provide an on-time delivery guaranteed by bounding the length of the MS path, which may be relaxed in the case of DTNs. Additionally, they leave out consideration of a disjoint and large-scale network as in DTNs. In such cases, the task of determining an appropriate number of sojourn locations and an optimal number of MSs becomes even more crucial to enhance the performance of the network. Most importantly, the propagation model in the existing schemes is assumed to be oversimplified as they do not follow the signal path-loss model. This should not be overlooked as the signal strength drops considerably when an entity moves away from its center, hampering the real-time data collection. Therefore, a realistic propagation model that imbibes the log-normal shadowing phenomena is overriding to design data collection schemes for real-life applications [56]. Apart from the above, none of them have considered the following issues. 1) How to detect the position of RPs in convex part of the obstacles and how many RPs (overall) will be there in the network, 2) How to re-route the MS for its seamless movement around the obstacle, and 3) How to remove sharp turns in the path of the MS. In other words, they have neglected smooth path construction, i.e., the path must be obstacle-avoiding that combines the properties of being both smooth and fast to traverse [84] so that delay and energy consumption will be significantly reduced. A summary of the selected hybrid data collection schemes evaluated on various parameters has been shown in Table 2.2.

The thesis is motivated by the shortcomings of the above-discussed works. In the subsequent chapters, we detail the works which remove the above-mentioned limitations and present novel and efficient schemes for data collection.

Chapter 3

A Heuristic Based Data Collection Scheme

Chapter 3

A Heuristic Based Data Collection Scheme

3.1 Introduction

In this chapter, we address the single-hop data collection problem using single MS using a heuristic approach. In particular, we optimize the number of rendezvous points and their locations to minimize the traveling length of the mobile sink. The scheme is based on spectral clustering (SC) and is thus called spectral clustering tour (SCTR). The rationale behind the use of SC is as follows: 1) it is able to achieve balanced clusters where SNs are distributed evenly, and this leads to the balancing of energy consumption of SNs, 2) it is able to provide a better quality of clustering as there is no chance to get trapped in local minima, 3) it converges faster as simple linear algebra is involved in the process and thus bears less computational overhead. In addition to this, unlike other clustering schemes, which make strong assumptions on the statistics of the clusters, the SC is easy to implement and helps to create more accurate clusters. In the proposed scheme, the connected components of the network are obtained using breadth-first search (BFS) traversal. Then, by jointly utilizing a heuristic and Tarjan's algorithm, the network is divided into subgraphs (i.e., partitions) which are referred to as manageable units (MUs). RPs are then determined from the MUs by fulfilling a threshold value. Finally, the RPs are used to form the traveling salesperson(TSP) tour for the MS using Christofide's algorithm [100]. The proposed method is extensively simulated and compared with three existing algorithms, namely spanning tree covering algorithm [36], CSS [101], and TSP in terms of path length, number of RPs, the variation of RPs, and energy consumption per round.

The simulation results prove the effectiveness of SCTR over the existing algorithms. The technical details of the proposed scheme and simulation study is elaborated in the subsequent sections. However, we begin by explaining the system model.

3.2 System Model

We consider a 2-D square area in which the SNs are deployed randomly which are assumed to be homogeneous, energy-constrained, and static after deployment. There is an MS which is responsible for data collection. The following assumptions are made for the MS. 1) It is aware of the location of the SNs and the BS. 2) Its communication range is sufficient to collect data from the SNs in one hop while at an RP. 3) It communicates the data of the SNs directly to the BS. 4) The speed is enough to cover the tour in stipulated time. Note that we consider the network as a graph $G = (S, E)$, where S is the set of SNs and E is the set of edges between SNs in S. An edge between two SN exists if they are in communication range of each other. Additionally, in our system arrangement, we consider the same energy model as in [102] as follows. In this model, there are two parts: free space (where power loss is proportional to d^2) and multipath fading (where power loss is proportional to d^4). If the distance is found to be less than the threshold value (d_0), then the free space (fs) model is used; otherwise, the multipath (mp) model is used. The energy needed by the radio hardware to transmit a k bit message over a distance d is ($E_{Tx}(k,d)$):

$$\begin{aligned} E_{Tx}(k,d) &= E_{Tx-elec}(k) + E_{Tx-amp}(k,d) \\ &= \begin{cases} kE_{elec} + k\xi_{fs}d^2 & \text{, if } d < d_0 \\ kE_{elec} + k\varepsilon_{mp}d^4 & \text{, if } d \geq d_0 \end{cases} \end{aligned} \tag{3.1}$$

where $E_{Tx-elec}$ is the energy spent by the transmitter electronics for transmission, E_{Tx-amp} is the energy spent by the amplifier circuit.

The energy spent by the radio to receive k bit message over a distance d is (E_{Rx}):

$$E_{Rx} = E_{Rx-elec}(k) = kE_{elec} \tag{3.2}$$

where $E_{Rx-elec}$ is the energy spent by the receiver electronics. The E_{elec} (electronics energy) depends on various factors such as digital coding, spreading of the signal, modulation, and filtering. The amplifier energy (either $\xi_{fs}d^2$ or $\varepsilon_{mp}d^4$) depends on two factors: distance from the transmitter to the receiver and acceptable bit-error rate. It is worth mentioning that this is highly simplified modelling of the actual radio wave

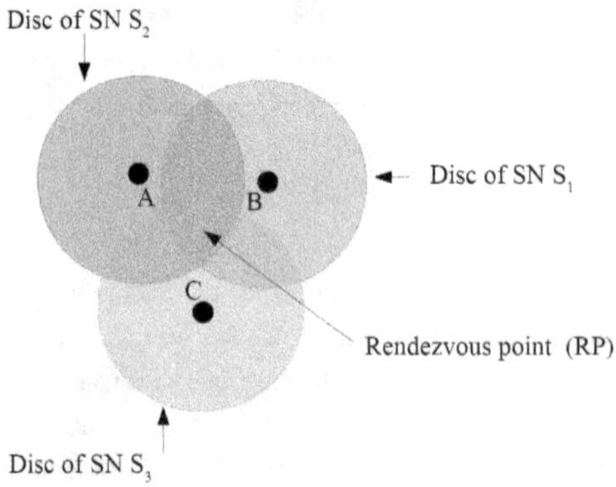

Figure 3.1: Rendezvous point of three SNs.

propagation, which is very unpredictable and also highly difficult to model.

3.2.1 Definitions and Symbols

Before we formally describe the data-gathering problem, we first define some definitions to understand the rest of this paper.

Definition 3.2.1. *: **Tour of the MS (D_{tour})***
A tour is defined as the path travelled by the MS to collect data from the SNs by visiting RPs in one hop communication. It starts and ends with the location of the BS (r_0), i.e.,

$$D_{tour} = \; < r_0, r_1, \ldots, r_{n^r}, r_0 > . \tag{3.3}$$

It's length is denoted by $|D_{tour}|$.

Definition 3.2.2. *: **Rendezvous point (RP)***
An RP is defined as a location at which the MS sojourns and collects data from the SNs in its communication range. The communication range of each SNs is assumed as a disk (disk Model [103]), and if there are several SNs in communication range of each other then their disks intersect and the centroid of the overlapping region is called as an RP.

In Fig. 3.1 three SNs s_1, s_2, and s_3 are shown along with their disc shaped communication range. These three discs have an equal radius. Their overlapping area is marked by A, B, and C and the centroid of this region is considered as the RP of SNs s_1, s_2, and s_3.

Definition 3.2.3. *: **$cut(A_1, A_2)$***
Let A_1,A_2 be two subgraphs of a graph G which is partitioned into two components by

removing edges between A_1 and A_2. Then, the dissimilarity between these two partitions can be measured using the weight of the edges that have been removed. This is called a cut between A_1 and A_2, which is mathematically expressed as follows:

$$cut(A_1,A_2) = \sum_{i \in A_1} \sum_{j \in A_2} d_{i,j} \tag{3.4}$$

where $d_{i,j}$ is the weight (euclidean distance) of edges between A_1 and A_2.

Definition 3.2.4. : ***$Ncut(A_1,A_2)$***
An $Ncut(A_1,A_2)$ is defined mathematically as follows.

$$Ncut(A_1,A_2) = cut(A_1,A_2)(\frac{1}{vol(A_1)} + \frac{1}{vol(A_2)}) \tag{3.5}$$

where

$$vol(A_k) = \sum_{i \in A_k} d_i \tag{3.6}$$

which is the sum of edge weights of all edges attached to $A_k (k = 1,2)$. Any cut partitions the network in which there may be one SN. So normalized cuts are used which ensures that the partitions are balanced. Now, extending Ncut for k number of disjoint partitions for a graph G, the normalized cut can be written as

$$Ncut(A_1,A_2,\ldots,A_k) = \frac{cut(A_1,A_1^{\complement})}{vol(A_1)} + \ldots + \frac{cut(A_k,A_k^{\complement})}{vol(A_k)}. \tag{3.7}$$

where $A_i^{\complement}$ denotes the compliment of A_i. (3.7) can be again rewritten in terms of indicator vector, degree matrix, and adjacency matrix as follows [104]:

$$Ncut(A_1,A_2,\ldots,A_k) = \frac{1}{2}\sum_{i=1}^{k} \frac{x_i^{I^T}(D_G - A_G)x_i^I}{x_i^{I^T} A_G x_i^I}. \tag{3.8}$$

where x_i^I is an indicator vector of mu_i, $x_i^{I^T}$ is the transpose of the indicator vector, D_G is the degree matrix of mu_i, A_G is the adjacency matrix of mu_i.

In order to divide a graph into k number of connected components we need to optimize the normalized cut as stated in (3.8).

Definition 3.2.5. : ***Lowest and highest of p_{val}***
In a directed graph as shown in Fig. 3.2, a positive number is assigned to each node (p_{val}) according to a post order traversal (black color). Now, for each s_i, the descendants of s_i including itself (red color) and those which are reachable from s_i using a back edge in one hop, the least p_{val} of all these nodes is considered as the lowest of p_{val}, and denoted

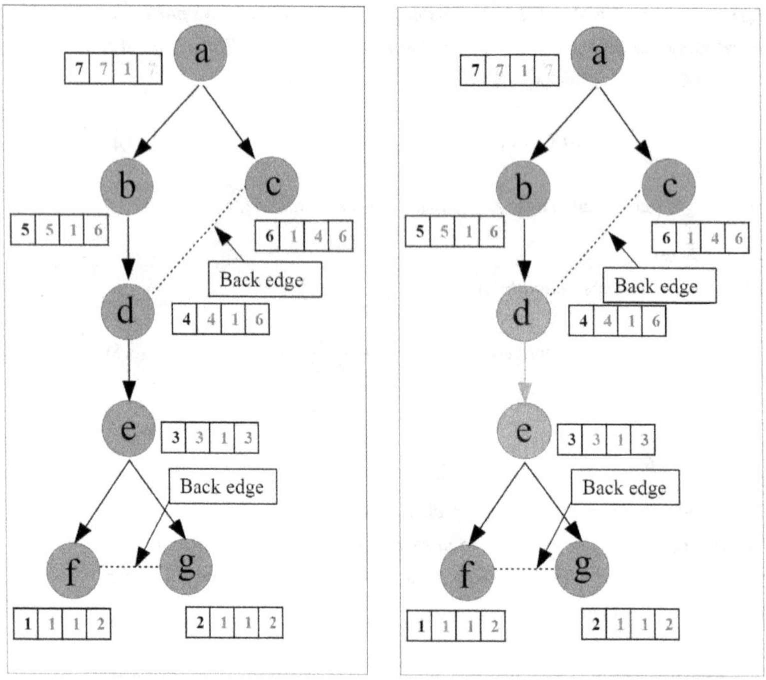

Figure 3.2: Finding bridge in a directed graph. The edge (d,e) is a bridge according to Tarjan's Algorithm.

by l_i. Similarly, the maximum p_{val} of all those SNs is considered as the highest of p_{val}, and denoted by h_i. The l_i and h_i of each SN s_i is shown in Fig. 3.2 with blue and green colors, respectively.

3.3 Proposed Work

The basic idea of our proposed scheme is given in the form of a flowchart as shown in Fig. 3.3. We explain each of these steps in the subsequent sections. However, let us first present the problem formulation as follows.

3.3.1 Problem Formulation

We consider the problem of finding the shortest path traversed by the MS to collect data from SNs in one hop. Let a_r be the radius of the deployment area and c^s be the

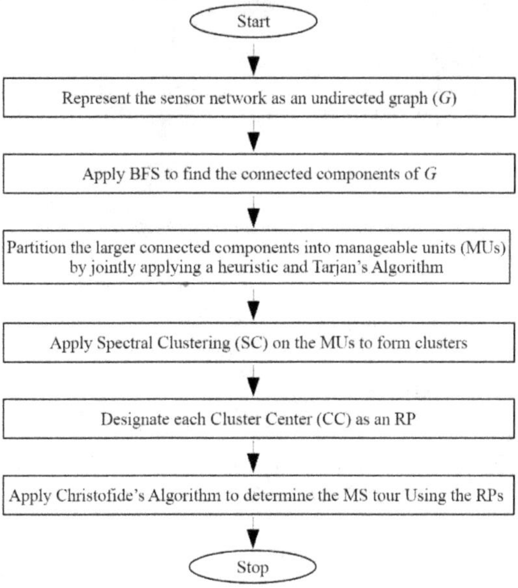

Figure 3.3: Flowchart to demonstrate the working procedure of the proposed algorithm.

communication range of an SN, then $\frac{c^s}{a_r} << 1$ and $0 < \frac{c^s}{a_r} < 1$. In this situation, the SNs have overlapping communication range and data from several SNs can be collected from a single location. All such locations can be assumed to be points, and data collection from such points can be mapped to a TSP problem, so it is considered to be an NP-hard problem.

The problem formulation has two parts. First, we formulate the problem of minimizing the normalized cut of all sufficiently large MUs as follows. Therefore, the first objective function can be stated mathematically as follows:

$$minimize(Ncut(A_1,A_2,\ldots,A_k)). \tag{3.9}$$

The second objective can be expressed as an integer linear programming as follows.

Let $b_{i,j}$, O_i, and $h_{i,j}$ be the Boolean variables defined as follows.

$$b_{i,j} = \begin{cases} 1 & \text{if } r_i \text{ and } r_j \in D_{tour}, \\ 0 & \text{otherwise} \end{cases} \tag{3.10}$$

,

$$O_i = \begin{cases} 1 & \text{if } r_i \in D_{tour}, \\ 0 & \text{otherwise} \end{cases} \tag{3.11}$$

and

$$h_{i,j} = \begin{cases} 1 & \text{if } s_i \text{ is covered in one hop distance by } r_j, \\ 0 & \text{otherwise.} \end{cases} \tag{3.12}$$

. Then, the path minimization problem is given by

$$minimize\ (D_{tour}) \tag{3.13}$$

where

$$D_{tour} = \sum_{1 \leq i,j \leq n^r, i \neq j} d_{i,j} b_{i,j} \tag{3.14}$$

subject to

$$\prod_{r_i \in R} O_i = 1, \tag{3.15}$$

$$\sum_{1 \leq i \leq n^r} n_{r_i} = S, \tag{3.16}$$

and

$$\prod_{1 \leq i \leq n^s, 1 \leq j \leq n^r} h_{i,j} = 1. \tag{3.17}$$

The constraint in (3.15) ensures that each of the RPs is included in the tour. It also indicates that each RP is visited only once in the tour. Constraint (3.16) ensures that all SNs are covered in the tour. Constraint (3.17) ensures that all SNs are sending data to the RPs in one hop transmission.

3.3.2 Connected Components Determination

The network is represented as an undirected graph (G) as explained in section 3.2. Then, connected components of the network are obtained by applying Algorithm 1, which involves BFS traversal (Algorithm 2). Line 1 of Algorithm 1 initializes a set with empty elements. Next, each SN is marked as not traversed (using *marked*) and initial component number set to zero (using *setCompNo*) in Lines 3-5. After that, each SN is traversed based on fulfilling the condition for BFS traversal. This is performed in Lines 6-10. At last, a

Algorithm 1: senNetComp: Algorithm to find connected components of G

```
Input  : G
Output: ψ
1  ψ = φ
2  for each vertex s_i in S of G do
3      marked(s_i) = 0
4      setCompNo(s_i) = 0
5  end
6  for each vertex s_i in S of G do
7      if marked(s_i) == 0 then
8          CompNo = CompNo + 1
9          BFS(G, s_i, CompNo)
10     end
11 end
12 ψ ← component number of all SNs
```

Algorithm 2: BFS

```
Input  : G, s, CompNo
Output: All SNs whose component number is CompNo and also is reachable
        from s
Q = φ                                  // Initializes an empty queue
Enqueue(Q, s)                          // Perform add operation on queue Q
marked(s) = 1
setCompNo(s) = 1
while Q is not empty do
    u = Dequeue(Q)             // Perform remove operation on queue Q
    for each vertex v in adjacent vertices of u do
        if marked(v) == 0 then
            marked(v) = 1
            setCompNo(v) = 1
            Enqueue(Q, v)
        end
    end
end
ψ ← component number of all SNs
```

set containing the component number of each SN (ψ) is obtained (Line 12). Algorithm 2 performs the traversal of the SNs to determine all SNs reachable from starting SN (root node (s)) and assigns a value *CompNo* to them.

Algorithm 3: findBridge:Algorithm to find Bridges

Input : ψ'
Output: B_r
$B_r = \phi$
$d_T \leftarrow$ build a DFS tree using DFS search of ψ'
Perform post order traversal of d_T and assign p_{val_i} to S_i
for *each SN S_i in ψ'* **do**
    $n_{d_i} \leftarrow$ number of descendants of node S_i
    $l_i \leftarrow$ lowest of p_{val_i} of descendants of node S_i
    $h_i \leftarrow$ highest of p_{val_i} of descendants of node S_i
end
for *each edge e in ψ'* **do**
    if *$h_i \leq p_{val_i}$ and $l_i > (p_{val_i} - n_{d_i})$* **then**
        $B_r = B_r \cup e$
    end
end

3.3.3 Manageable Units (MUs) Formation

We divide the network into MUs such that each MU is a connected subgraph of G. The motivation behind forming MUs is to reduce the problem size (note that graph partitioning problem is NP hard in nature) that would reduce the computational complexity of the solution.

The pseudocode to form MU is given in Algorithm 4. It requires the connected components of the network, which are obtained with the help of BFS traversal algorithm (Line 3 of Algorithm 4). For each of the connected components, bridges are found using the Tarjan's algorithm (refer Algorithm 3). Note that this algorithm works for directed graphs, but the connected components involved here are undirected graphs, so at first we convert them into directed graphs to obtain bridges. For each of these directed graphs, all bridge are obtained using Algorithm 3. After obtaining the bridges of each connected component, they are selectively removed and checked, if removing a bridge subdivides the component such that its size is less than k_0, then it is added back, otherwise two more components are obtained. Likewise, all connected components are converted into subgraphs of suitable size known as manageable units.

We take the value of $k_0 = 3$. The reason is, if the value of $k_0 > 3$, the components will have a sufficient number of SNs to partition and form clusters. If the value of $k_0 < 3$, then the components does not need partition because they can be treated as an MU. This is performed in the Lines 4-10 of Algorithm 4. The outcome of the Algorithm 4 is a set of MUs (μ^{mu}).

3.3.4 Cluster Formation and Path Determination

Once the MU formation phase is complete, RP formation process begins. The basic idea here is to partition the graph by SC and find the centroid of each division to form RPs. These RPs are finally used to plan a tour for the MS. For SC, at first laplacian matrix is used which is defined as follows.

$$L = D_G - A_G \tag{3.18}$$

where D_G is the degree matrix and A_G is the adjacency matrix. We then define the degree matrix $D_G = d_{G_{i,j}}$ as follows:

$$d_{G_{i,j}} = \begin{cases} 0 & \text{if } i \neq j, \\ deg(s_i) & \text{if } i == j \end{cases} \tag{3.19}$$

where $deg(s_i)$ is the degree of node s_i. The adjacency matrix $A_G = a_{i,j}$ is defined as follows:

$$a_{i,j} = \begin{cases} 0 & \text{if no edge exists between } s_i \text{ and } s_j, \\ 1 & \text{if edge exists between } s_i \text{ and } s_j. \end{cases} \tag{3.20}$$

The pseudocode of SCTR is given in Algorithm 5. In Line 1, Algorithm 4 is called to form MUs of the graph G. Now, SC is performed for all those MUs whose size is greater than k_0. It requires specifying a maximum value of k, which is estimated using elbow method (Lines 4-7). Line 8 builds the adjacency matrix and Line 9 forms the degree matrix of the MU. In Line 10, laplacian matrix is obtained using (3.18). In Lines 11-12, k number of eigenvectors are calculated and stored in U. In Line number 13, each row of U is stored in a matrix Z. Now in Line 14, these points are clustered using k-means algorithm and the resulting cluster centres are stored in r_p. In a similar manner, this process is repeated for all the MUs that satisfies the previously mentioned condition. For MUs, whose size is less than k_0, SCTR is not applied and the centroid of the points is considered as RP. This is performed in Lines 16-19. Finally, TSP algorithm is applied to r_p to form a tour for MS in step 20.

3.3.5 Time Complexity

In this section, we analyze the time complexity of the proposed algorithms. In Algorithm 1, the connected components are determined using BFS traversal (Algorithm 2). Here, Algorithm 2 takes $O(n^s + e)$ time, where n^s is the number of SNs and e is the number of edges in that component. Assuming, there are k^1 number of connected

Algorithm 4: MuForm: Algorithm to form MUs

Input : $G(S,E)$
Output: μ^{mu}

$B_r = \phi$
$\psi \leftarrow$ find connected components of G using Algorithm 1
for *each set* ψ_i *in* ψ **do**
  $B_{r_i} \leftarrow$ FINDBRIDGE(ψ_i)
  for *each bridge edge br in* B_{r_i} **do**
    $C_{p_1}, C_{p_2} \leftarrow$ remove *br* from B_{r_i}
    if $|C_{p_1}| or |C_{p_2}| <= k_0$ **then**
      Add *br* back to B_{r_i}
    end
  end
end
$E \leftarrow E - B_r$ // B_r is the set of bridge edges
$\mu^{mu} \leftarrow$ find the connected components of G using Algorithm 1

Algorithm 5: SpecCluTour: Algorithm to find RPs and plan a tour for data collection (SCTR)

Input : G
Output: D_{tour}

$\mu^{mu} \leftarrow MuForm(G)$
$rp \leftarrow \phi$
for *each MU* mu_i *in* μ^{mu} **do**
  if $|mu_i| > k_0$ **then**
    Build clusters using k-means algorithm with k, $1 \leq k \leq |mu_i|/2$
    Find total within-cluster sum of square each k
    Plot graph between k and total within-cluster sum of square
    $k \leftarrow$ location of the bend (elbow) in the plot
    $A_{mu_i} \leftarrow$ build adjacency matrix for nodes in mu_i
    $D_{mu_i} \leftarrow$ build degree matrix for nodes in mu_i
    $L \leftarrow D_{mu_i} - A_{mu_i}$
    Find the first k eigenvectors $u_{e_1}, u_{e_2}, \ldots, u_{e_k}$ of L
    $U \leftarrow [u_{e_1}, u_{e_2}, \ldots, u_{e_k}], U \in \mathbb{R}^{|mu_i| \times k}$
    $Z_j \leftarrow j^{th}$ row of $U, Z_j \in \mathbb{R}^k$
    $r_p \leftarrow r_p \cup$ Cluster the points Z_j using k-means algorithm
  end
  if $|mu_i| \leq k_0$ **then**
    $r_p \leftarrow r_p \cup$ centroid of mu_i
  end
end
$D_{tour} \leftarrow$ Apply TSP Algorithm to r_p to form a tour for the MS using Christofide's heuristic [100]

components Algorithm 1 calls Algorithm 2, k^1 times and hence requires $O(k^1(n^s+e))$ time. Therefore, total time complexity of Algorithm 1 is $O(n^s+e)$.

For Algorithm 3, Line 1 takes $O(1)$ time and Line 2 takes $O(n'+e')$ time considering adjacency list to represent the graph where n' is the number of nodes in the sub-graph and e' is the number of edges in the sub-graph. Line 3 takes $O(n')$ time. Each of the Lines 5, 6, and 7 take $O(n')$ time. So Lines 4-8 take $O(n'^2)$ time. Lines 9-13 take $O(e')$ time. Therefore, total time complexity of Algorithm 3 is $O(n'^2)$. In Algorithm 4, Line 2 uses Algorithm 1 so the time complexity is $O(n^s+e)$, where n^s is the number of SNs and e is the number of edges in the graph. Let k' be the number of connected components, then Lines 3-11 take $O(k'((n^s)^2))$ time. After finding and removing all the bridge edges, connected components are again found which takes $O((n^s)^2)$ time (Lines 12-13). So, total time complexity is $O((n^s)^2)$.

For Algorithm 5, Line 1 takes $O((n^s)^2)$ time. Let there are t' number of MUs then average number of SNs per MU is $\frac{n^s}{t'}$, so Lines 5-6 takes $O(\frac{n^s}{t'}kt)$ time, where t is the number of iterations. Lines 7-8 take $O(1)$ time. Lines 8-9 take $O(\frac{n}{t'}^2)$ time. Line 11 also takes $O(\frac{n^s}{t'}^2)$ time. Lines 13-15 take $O(k^3)+O(k\frac{n^s}{t'}t)$ time [105]. Lines 17-19 take $O(1)$ time. Hence total time complexity of Lines 3-20 is $O((n^s)^2)$. It is to be noted that Line 21 is a call to the TSP algorithm which is solved using Christofide's heuristic [100], therefore total time complexity of SCTR algorithm is $O((n^s)^2)+O((n^s)^3) \approx O((n^s)^3)$.

3.3.6 A Numerical Example

We consider a sensor network as shown in Fig. 3.4(a). There are 12 sensor nodes in the network. The base station is shown by the *cyan* color and is placed in the centre of the network. In Fig. 3.4(b) we show the SNs which are in the communication range of each other indicated by the edges between them. At first, the connected components of the network is found using BFS traversal. The output of the algorithm is as follows:

$$\psi_1 : \{1,2,3,4,5,6,7,8,9\}, \psi_2 : \{10,12\}, \psi_3 : \{11\}.$$

Since there are no bridges in any of the connected components, after applying Algorithm 4 only three MUs are returned. The output of this step is as follows. $mu_1 = (V_1, E_1)$, $mu_2 = (V_2, E_2)$, $mu_3 = (V_3, E_3)$, where $V_1 = \{1,2,3,4,5,6,7,8,9\}$, $E_1 = \{(1,2), (1,3), (1,5), (2,4), (2,5), (3,5), (4,7), (4,9), (5,4), (5,7), (6,8), (6,7), (6,9), (7,9), (7,8), (8,9)\}$, $V_2 = \{10,12\}$, $E_2 = \{(10,12)\}$, $V_3 = \{11\}$, $E_3 = \{\}$. Here, size of MUs mu_2 and mu_3 are less than 3, so SC is not applied to these MUs and the centroid of the location of the SNs are designated as RPs. For mu_2 and mu_3 the location of RP is indicated as R_2 and R_3, respectively as shown in Fig. 3.4(c). The size of MU (mu_1)

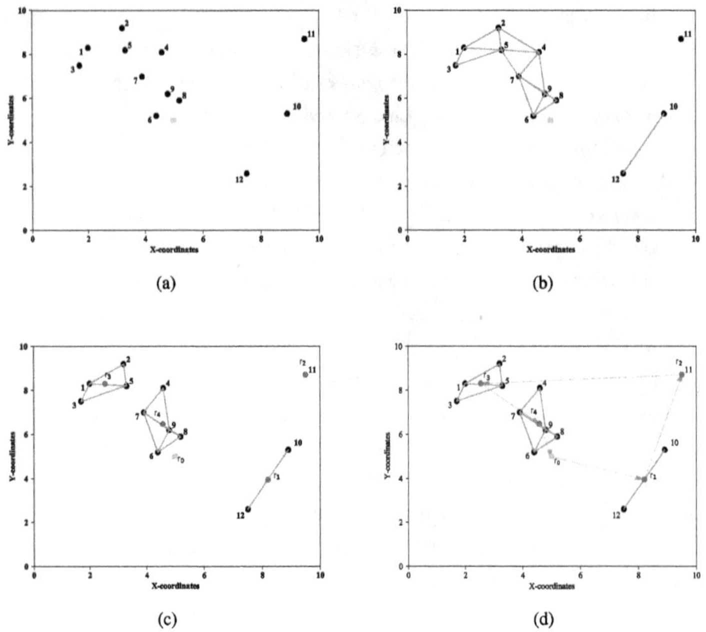

Figure 3.4: An illustration of the proposed algorithm. (a) Random Deployment of SNs. (b) Graphical representation of network. (d) MUs of the network. (d) Path of the MS.

is greater than k_0 so SC is applied to it. The adjacency matrix A_{mu_1} for MU mu_1 is calculated as follows.

$$A_{G_1} = \begin{pmatrix} 1 & 1 & 1 & 0 & 1 & 0 & 0 & 0 & 0 \\ 1 & 1 & 0 & 1 & 1 & 0 & 0 & 0 & 0 \\ 1 & 0 & 1 & 0 & 1 & 0 & 0 & 0 & 0 \\ 0 & 1 & 0 & 1 & 1 & 0 & 1 & 0 & 1 \\ 1 & 1 & 1 & 1 & 1 & 0 & 1 & 0 & 0 \\ 0 & 0 & 0 & 0 & 0 & 1 & 1 & 1 & 1 \\ 0 & 0 & 0 & 1 & 1 & 1 & 1 & 1 & 1 \\ 0 & 0 & 0 & 0 & 0 & 1 & 1 & 1 & 1 \\ 0 & 0 & 0 & 1 & 0 & 1 & 1 & 1 & 1 \end{pmatrix} \tag{3.21}$$

The degree matrix D_{mu_1} and the laplacian matrix L is calculated as follows.

$$D_{G_1} = \begin{pmatrix} 3 & 0 & 0 & 0 & 0 & 0 & 0 & 0 & 0 \\ 0 & 3 & 0 & 0 & 0 & 0 & 0 & 0 & 0 \\ 0 & 0 & 2 & 0 & 0 & 0 & 0 & 0 & 0 \\ 0 & 0 & 0 & 4 & 0 & 0 & 0 & 0 & 0 \\ 0 & 0 & 0 & 0 & 5 & 0 & 0 & 0 & 0 \\ 0 & 0 & 0 & 0 & 0 & 3 & 0 & 0 & 0 \\ 0 & 0 & 0 & 0 & 0 & 0 & 5 & 0 & 0 \\ 0 & 0 & 0 & 0 & 0 & 0 & 0 & 3 & 0 \\ 0 & 0 & 0 & 0 & 0 & 0 & 0 & 0 & 4 \end{pmatrix} \tag{3.22}$$

$$L = \begin{pmatrix} 2 & -1 & -1 & 0 & -1 & 0 & 0 & 0 & 0 \\ -1 & 2 & 0 & -1 & -1 & 0 & 0 & 0 & 0 \\ -1 & 0 & 1 & 0 & -1 & 0 & 0 & 0 & 0 \\ 0 & -1 & 0 & 3 & -1 & 0 & -1 & 0 & -1 \\ -1 & -1 & -1 & -1 & 4 & 0 & -1 & 0 & 0 \\ 0 & 0 & 0 & 0 & 0 & 2 & -1 & -1 & -1 \\ 0 & 0 & 0 & -1 & -1 & -1 & 4 & -1 & -1 \\ 0 & 0 & 0 & 0 & 0 & -1 & -1 & 2 & -1 \\ 0 & 0 & 0 & -1 & 0 & -1 & -1 & -1 & 3 \end{pmatrix} \tag{3.23}$$

The eigenvalues of the laplacian matrix are obtained as follows.

$$\{-1.000, -0.311, 1.151, 2.394, 3.000, 3.351, 3.944, 4.889, 5.582\} \tag{3.24}$$

The eigenvectors corresponding to each of the eigenvalues are as follows.

$$\begin{bmatrix} 0.333 & 0.414 & -0.003 & -0.568 & 2.505e-17 & -0.595 & 0.0409 & -0.164 & -0.111 \\ 0.333 & 0.264 & 0.563 & -0.223 & 7.708e-16 & 0.581 & 0.252 & 0.168 & -0.153 \\ 0.333 & 0.479 & -0.669 & 0.309 & -1.136e-16 & 0.275 & 0.149 & -0.049 & -0.123 \\ 0.333 & -0.017 & 0.377 & 0.518 & 4.243e-17 & -0.137 & -0.050 & -0.675 & -0.014 \\ 0.333 & 0.214 & 0.104 & 0.138 & 9.837e-17 & -0.053 & -0.481 & 0.354 & 0.673 \\ 0.333 & -0.409 & -0.201 & -0.287 & -0.707 & 0.204 & -0.094 & -0.194 & 0.105 \\ 0.333 & -0.218 & 0.0284 & 0.185 & -4.017e-16 & -0.158 & -0.419 & 0.405 & -0.664 \\ 0.333 & -0.409 & -0.201 & -0.287 & 0.707 & 0.204 & -0.094 & -0.194 & 0.105 \\ 0.333 & -0.318 & 0.002 & 0.214 & -1.473e-16 & -0.321 & 0.696 & 0.348 & 0.181 \end{bmatrix} \tag{3.25}$$

Assuming that the optimal value of k is 2, we perform bisection of mu_1. Then

$$U = \begin{bmatrix} \mathbf{0.333} & \mathbf{0.414} \\ \mathbf{0.333} & \mathbf{0.264} \\ \mathbf{0.333} & \mathbf{0.479} \\ 0.333 & -0.017 \\ \mathbf{0.333} & \mathbf{0.214} \\ 0.333 & -0.409 \\ 0.333 & -0.218 \\ 0.333 & -0.409 \\ 0.333 & -0.318 \end{bmatrix} \tag{3.26}$$

Now, each row of U denoted as Z_i is clustered using k-means algorithm. This way we obtain two clusters from mu_1. The centroid of the two clusters are marked as r_3 and r_4, respectively as shown in Fig. 3.4(c). Finally, RPs r_1, r_2, r_3, and r_4 are used to plan the tour for the MS. As shown in the Fig. 3.4(d), the path of the MS starts from the location of the base station (r_0) and after data collection ends at the same location.

3.4 Performance Comparisons

In this section, we present the simulation study of our proposed work. We compare our algorithm with spanning tree covering based algorithm [36], CSS [101], and TSP algorithm where one hop delivery of data occurs. We consider different parameters for comparison such as path length, number of rendezvous points, the variation of RPs, and energy consumption of SNs in one round of the tour. In the simulation experiments, the parameter settings are as follows. The size of the network is varied from $200 \times 200\ m^2$ to $600 \times 600\ m^2$. The number of SNs is varied from 50 to 350. The initial energy of an SN is set to 30 J. The communication range of an SN is 30 m and 40 m. The simulation is performed in MATLAB 2017a on Ubuntu operating system (64 bit). The processor is Intel(R) Core(TM) i7-7500/ CPU @ 2.7 GHz $\times 4$ and 8 GB RAM.

3.4.1 Comparison Based on Rendezvous Points

Fig. 3.5 shows the comparison of the number of rendezvous points with other existing algorithms. The reason for adopting this parameter is that fewer RPs will result in lower computational cost as data collection is performed in single hop at each RP. It can be observed from Fig. 3.5 that TSP and CSS algorithms require large number of RPs, whereas the spanning tree covering and SCTR algorithms require much less number of RPs. This is attributed to the efficient organization of the network topology in spanning

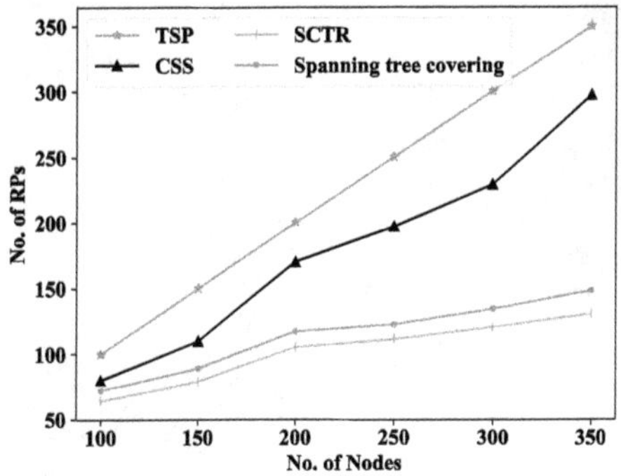

(a)

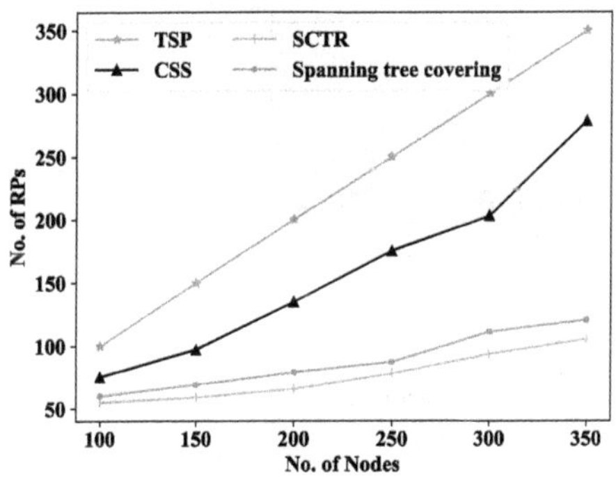

(b)

Figure 3.5: Comparison in terms of number of rendezvous points (a) $c^s = 30\ m$. (b) $c^s = 40\ m$.

tree covering and SCTR algorithms. The proposed algorithm was able to achieve 12% improvement over the spanning tree covering algorithm. Fig. 3.6 shows the effect of varying network size with the number of RPs for communication range 30 meters and 40 meters of SNs. It can be observed that the number of RPs increases as the network size increases due to the need for more RPs to cover the network area.

3.4.2 Comparison Based on Path Length

Fig. 3.7 shows the comparison of path length of the MS with other existing algorithms. It can be observed that the TSP and CSS algorithms have significantly longer path lengths,

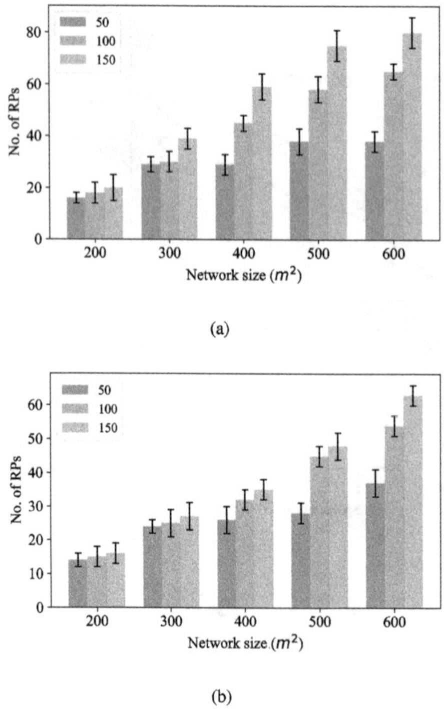

Figure 3.6: Variation of no. of RPs with respect to network size. (a) $c^s = 30\ m$. (b) $c^s = 40\ m$.

while the spanning tree covering algorithm and the SCTR algorithms have much shorter path lengths. Note that the path length becomes longer with the increase in the number of SNs. The reason is that for single hop data collection using MS fewer RPs is more likely to lead to a shorter path which occurs in the case of SCTR and spanning tree covering algorithm. The SCTR algorithm was able to achieve 10% improvement in path length over the spanning tree covering algorithm. Fig. 3.8 shows the effect of network size on the path length for communication range 30 meters and 40 meters of SNs. Note that the path length increases with the increase in network size, regardless of the number of SNs. The reason is that the SNs become more sparsely distributed with the increase in network size, and more RPs are required to cover the network area.

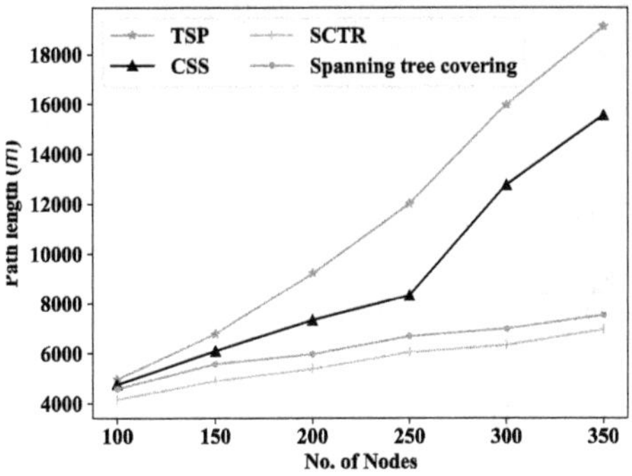

(a)

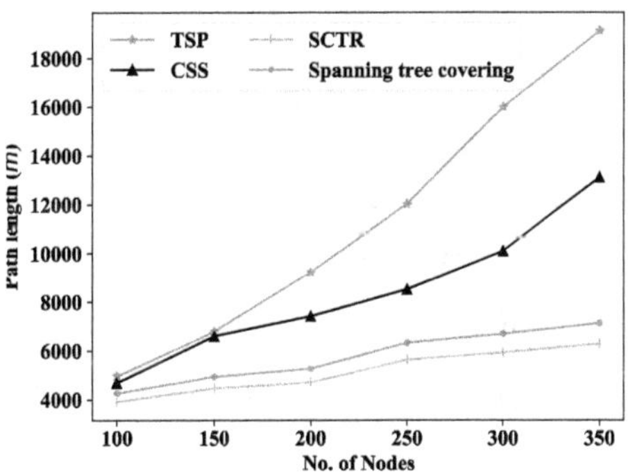

(b)

Figure 3.7: Comparison in terms of path length. (a) $c^s = 30\ m$. (b) $c^s = 40\ m$.

3.4.3 Comparison Based on Energy Consumption per Round

Fig. 3.9 shows the comparison of energy consumption per round with other existing algorithms. The reason behind considering this parameter is that less energy consumption per round would indicate that the network would remain operational for a longer period of time. It can be observed from Fig. 3.9 that the energy consumption of TSP and CSS algorithm is relatively high which can be attributed to the longer path length, on the other hand SCTR and spanning tree covering algorithm consumes a relatively small amount of energy due to relatively shorter path length. SCTR was able to achieve 30 %

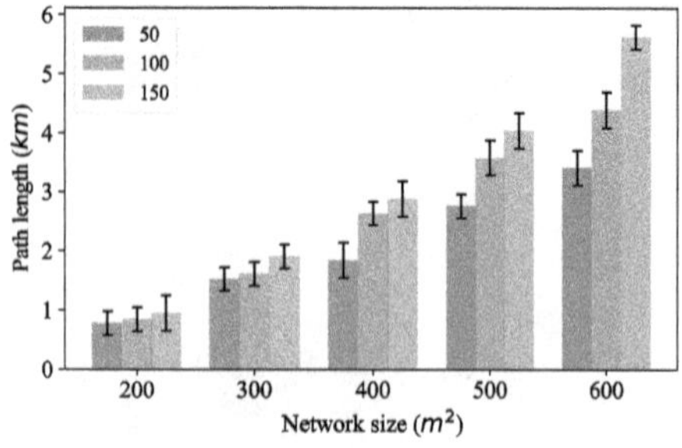

(a)

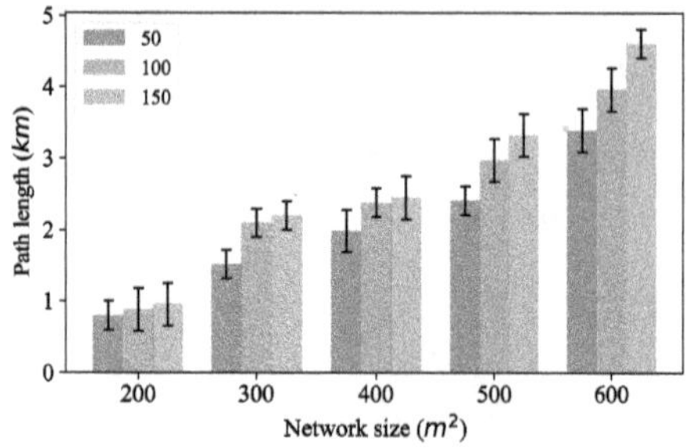

(b)

Figure 3.8: Comparison in terms of path length. (a) $c^s = 30\ m$. (b) $c^s = 40\ m$.

improvement over spanning tree covering algorithm for energy consumption per round. Fig. 3.10 shows the effect of variation of network size on energy consumption per round. Note that with the increase in network size, the amount of energy consumption per round increases linearly, regardless of the number of SNs.

3.4.4 Comparison Based on Variance of RPs

Fig. 3.11 shows the variance of RPs with other existing algorithms. Lower variation of RPs indicates that the number of SNs associated with RPs are balanced (which helps to reduce the buffer overflow problem of SNs). As shown in Fig. 3.11, our proposed algorithm has less variation of RPs as compared to the other algorithms. The SCTR algorithm was able to achieve 24 % improvement over the spanning tree covering algorithm. Note that we have not considered TSP algorithm in this comparison as the

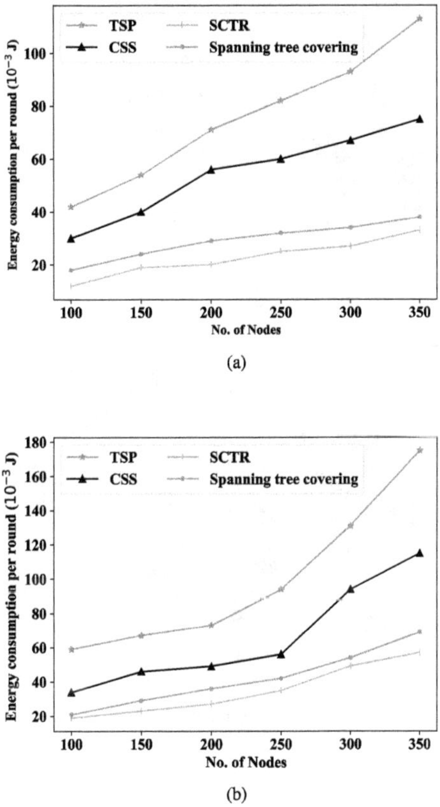

Figure 3.9: Comparison in terms of energy consumption per round (a) $c^s = 30\ m$. (b) $c^s = 40\ m$.

variation, in this case, will be zero because each SN is treated as an RP.

3.4.5 Statistical Analysis

In this section, we perform t-test [106] on the sample data generated for the spanning tree covering (STC) algorithm and the SCTR algorithm. It is a well known statistical test which determines whether two groups of data are significantly different from each other. For the test, we have considered two parameters, first is the number of RPs and the other is the path length. The number of SNs is considered to be 300, network size to be $600 \times 600\ m^2$, and the sample size is 15 for each group of data. According to the test, we

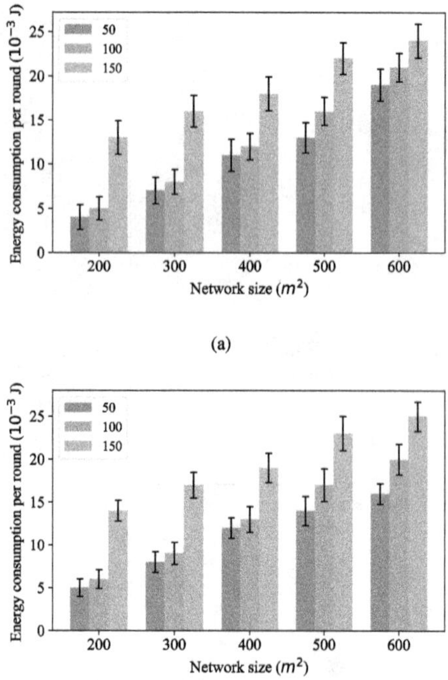

(a)

(b)

Figure 3.10: Comparison in terms of energy consumption per round (a) $c^s = 30\ m$. (b) $c^s = 40\ m$.

consider the null hypothesis as

$$H_0 : \mu_{STC} = \mu_{SCTR} \tag{3.27}$$

and the alternate hypothesis was considered as

$$H_1 : \mu_{STC} \neq \mu_{SCTR} \tag{3.28}$$

The null hypothesis (H_0) signifies that the means of the group are equal, whereas the alternate hypothesis indicates that the means of the group are significantly different. The null hypothesis (H_0) is accepted if p-value obtained using t-test is less than the α-value (0.05), otherwise the null hypothesis is accepted. It can be observed that in all cases

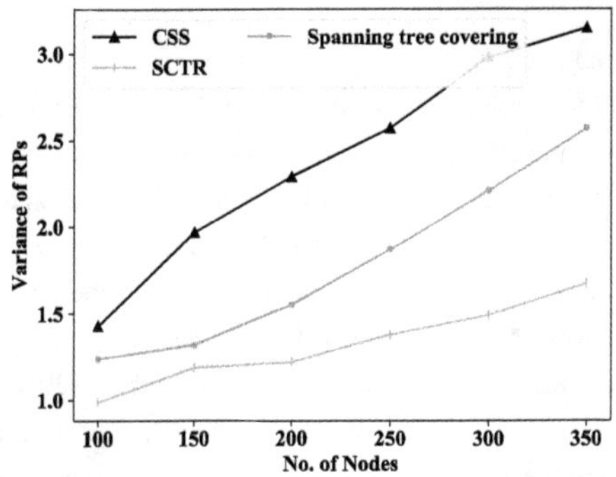

(a)

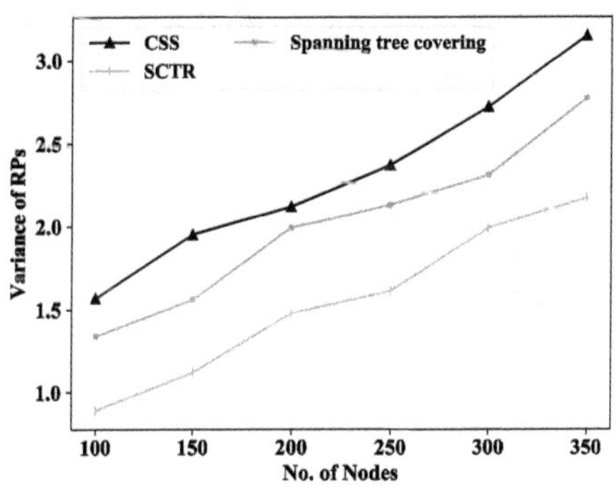

(b)

Figure 3.11: Comparison in terms of variance of RPs (a) $c^s = 30\ m$. (b) $c^s = 40\ m$.

Table 3.1: p-values produced by t-test between STC and SCTR algorithm

Range (m)	30		40	
Parameter	No. of RPs	Path length	No. of RPs	Path length
Test Statistic	-4.879	-7.357	-23.034	-14.081
p-value	3.862e-05	5.195e-08	9.689e-20	3.138e-14

the p-value obtained is much less than the α-value. So, the alternate hypothesis (H_1) is accepted and we can conclude that results obtained using SCTR algorithm are statistically significant and different from the spanning tree covering algorithm.

3.5 Chapter Summary

In this chapter, we have considered the tour planning problem by employing a single MS for data collection in the WSNs. Herein, we have first reduced the problem size by dividing the network into MUs by using BFS traversal and Tarjan's algorithm. Then, to obtain the optimal number and location of RPs, the SC algorithm is applied to each of the sufficiently large MUs. The proposed algorithm is shown to have a time complexity of $O((n^s)^3)$. Extensive simulations have validated the performance gain of the proposed approach over the competing approaches. The results have witnessed the proposed scheme to be promising by having a significant reduction in terms of number of RPs, path length, variation of RPs, and energy consumption per round. The validation of simulation results has also been shown through the statistical t-test. However, the scheme suffers in terms of performance in large-scale WSNs since a single MS may not be sufficient to collect data from all the nodes within the desired time frame. Thus, we investigate the use of multiple MSs in the next chapter using a recent metaheuristic algorithm called the shark smell optimization algorithm.

Chapter 4

Shark Smell Based Data Collection Scheme

Chapter 4

Shark Smell Based Data Collection Scheme

4.1 Introduction

In this chapter, we develop a novel single hop-based data collection scheme for large and disjoint networks. The problem considerd here is to find the optimal set of RPs (number, location, and sequence) and optimal number of MSs. The scheme to solve the problem is composed of two algorithms that are based on a recently developed nature-inspired scheme called shark smell optimization (SSO) technique [63] which is shown to be more advantageous in comparison to the conventional metaheuristic algorithms such as GA and PSO, because of the following facts: 1) the local search is exhaustive, providing a better locally optimal solution, and 2) the accuracy and convergence rate of the search is higher due to the use of a gradient function. Different from the existing schemes, the proposed method jointly considers the optimization of the number of RPs and the number of MSs for disjoint WSNs. We apply the first algorithm to find out an optimal number of RPs and their locations by representing it as a set-covering problem. The resulting RPs and their locations are then fed to the second algorithm to determine the optimal number of MSs by treating it as a subtour optimization problem for finding the optimal traveling paths. To the best knowledge so far, we are the first to deal with both the above-mentioned optimization problems for large and disjoint networks. At first the overview of shark smell optimization is presented, and then the system model, problem statement, proposed work, and simulation analysis are presented in the subsequent sections.

4.2 Preliminaries

4.2.1 Shark Smell Optimization

The SSO is a recently developed metaheuristic algorithm [63] based on how a shark hunts its prey, more specifically, a fish. A shark has a powerful sense of smell, which makes it a superior hunter in the sea. Also, it can detect odour particles from a considerably distant location. The odour particles (OPs) are spilled by an injured fish's blood, which is pursued by a shark. It performs two different kinds of movements to reach the fish, forward and rotational motion.

The algorithm consists of a population of size n_p of OPs, with each OP having dimension n_d. An odour particle (OP) represents a complete solution, and the injured fish represents the optimal solution. Each OP has two components, position $(X_i^k, 1 \leq i \leq n_p)$ and velocity $(V_i^k, 1 \leq i \leq n_p)$. The superscript $k, 1 \leq k \leq k_{max}$ denotes the stage number. We use the following notation to represent the position vectors for the first stage of the population of OPs.

$$[X_1^1, X_2^1, \ldots, X_{n_p}^1] \tag{4.1}$$

where the components of the ith position vector in each dimension is represented as follows:

$$X_i^1 = [x_{i,1}^1, x_{i,2}^1, \ldots, x_{i,n_d}^1], 1 \leq i \leq n_p \tag{4.2}$$

The velocity of the OPs corresponding to the position vector is represented as follows:

$$[V_1^1, V_2^1, \ldots, V_{n_p}^1] \tag{4.3}$$

where the components of the ith velocity vector in each dimension is represented as follows:

$$V_i^1 = [v_{i,1}^1, v_{i,2}^1, \ldots, v_{i,n_d}^1], 1 \leq i \leq n_p \tag{4.4}$$

Each OP is evaluated by an objective function (OF) to judge the quality of the solution to the problem. In order to reach the optimal global solution, the algorithm follows forward movement and rotational movement. The search proceeds in a number of stages (k), wherein both the movements are performed. The forward movement is mathematically modelled as follows. At first, the velocity of the shark in k^{th} stage is updated using the following equation.

$$|v_{i,j}^k| = min\left[\left|\eta_k R_1' \frac{\partial(OF)}{\partial(x_j)}\right|_{x_{i,j}^k} + \alpha_k R_2' v_{i,j}^{k-1}\right|, \left|\beta_k v_{i,j}^{k-1}\right|\right], \quad 1 \leq i \leq n_p, 1 \leq j \leq n_d, 1 \leq k \leq k_{max} \tag{4.5}$$

where η_k is a constant belonging to $[0,1]$ also known as inertia coefficient or momentum rate, R_1' and R_2' are random numbers in the interval $[0,1]$, β_k is a constant which limits the current velocity of the shark such that it does not go beyond a certain limit than its previous velocity. Note that Eq.4.5 uses a gradient function, which improves the convergence rate of the algorithm. After that, the new position in k^{th} stage (Y_i^{k+1}) is determined using the following equation.

$$Y_i^{k+1} = X_i^k + V_i^k \Delta t_k, \qquad 1 \leq i \leq n_p, 1 \leq k \leq k_{max} \tag{4.6}$$

where Δt_k is time duration for k^{th} stage and is assumed to be 1 for all stages.

Now, the rotational movement is performed as follows. The algorithm selects some M additional points ($Z_i^{k+1,m'}, 1 \leq i \leq n_p, 1 \leq k \leq k_{max}, 1 \leq m' \leq M$) from the local search using the following equation.

$$Z_i^{k+1,m'} = Y_i^{k+1} + R_3' Y_i^{k+1}, \qquad 1 \leq i \leq n_p, 1 \leq k \leq k_{max}, 1 \leq m' \leq M \tag{4.7}$$

where R_3' is a random number belonging to $[-1,+1]$. In these points, if the shark finds a position with high odour intensity (i.e., locally optimal solution), it moves to that point and starts its global search. Finally, the OP is updated as follows:

$$X_i^{k+1} = arg\ max\ \{OF(Y_i^{k+1}), OF(Z_i^{k+1,1}), OF(Z_i^{k+1,2}), \ldots, OF(Z_i^{k+1,M})\}, 1 \leq i \leq n_p \tag{4.8}$$

The movement types, as mentioned above, are iteratively repeated until a termination criterion is reached. The flowchart of the SSO algorithm is given in Fig. 4.1. Note that the algorithm made the following assumptions. 1) The velocity of the injured fish is negligible as compared to a shark's velocity. 2) Blood from the fish is injected regularly, so the distortion in OPs due to water is negligible, and 3) There is only one blood source in the shark's search space, which is also the best optimal solution.

The SSO algorithm has the following advantages in comparison to the other population-based metaheuristic (e.g., PSO). 1) The local search is thorough due to the rotational movement of the shark. 2) There is a balance between local search (guided by rotational motion) and global search (guided by forwarding motion). 3) The convergence rate of the algorithm is high due to the use of a gradient function. These factors contribute

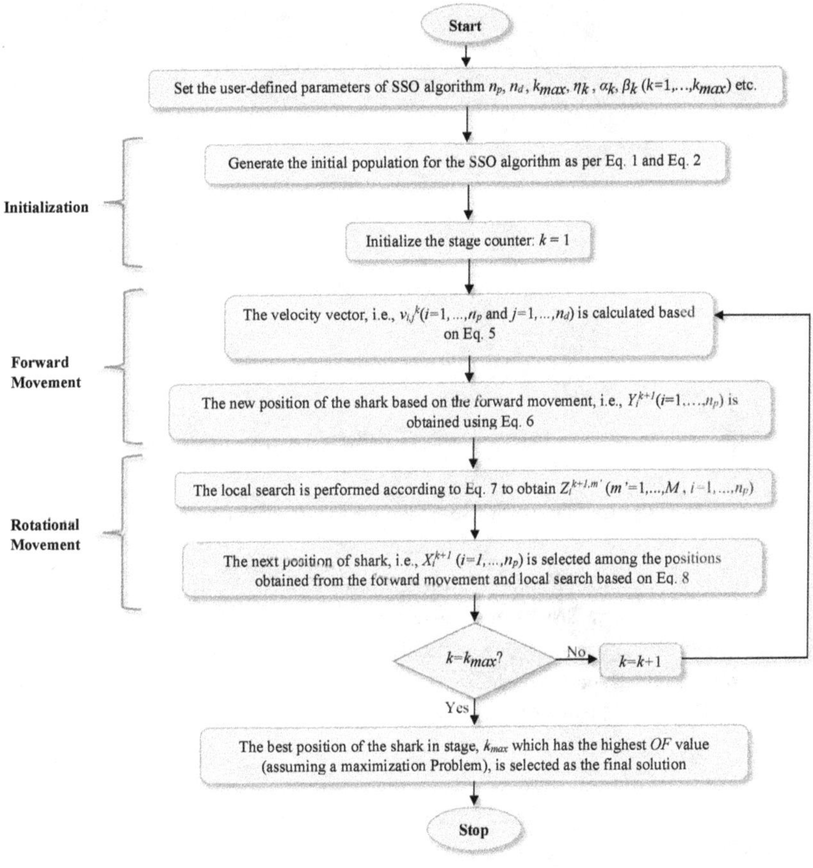

Figure 4.1: Flowchart of the SSO algorithm.

to obtaining an optimal solution efficiently and is the prime motivation behind using the SSO algorithm in this paper.

4.3 Network Environment

We consider a WSN comprising of a set of static SNs deployed over a 2-D area of size $l \times l$. Also, the network is segmented, i.e., we consider a disjoint WSN. Each MS knows the location of all SNs. They travel through the network according to the proposed algorithm for data collection from all SNs in the region. The MS is a resource rich device having

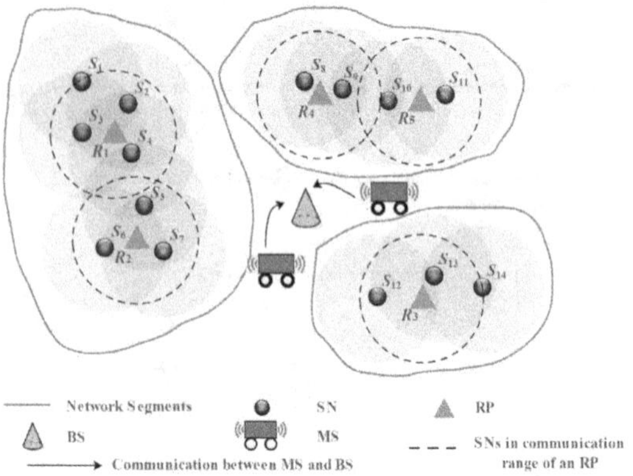

Figure 4.2: Network architecture for disjoint WSNs. (Refer S_i as s_i and R_i as r_i)

enough energy to complete the tour and it's speed is sufficient to complete the tour on time.

Fig. 4.2 shows an example of network architecture for disjoint WSN. There are three network segments where SNs of one segment are isolated from those of other segments. However, SNs in one segment are strongly connected with SNs of the same segment. Herein, the BS is located in the center of the network. There are 14 SNs, $S = \{s_1, s_2, \dots, s_{14}\}$ organised into 5 RPs, $R = \{r_1, r_2, \dots, r_5\}$. There are two MSs in the network.

4.3.1 Energy Model

We use the same energy model, as discussed in [102]. It is expressed mathematically as follows.

$$E_{Tx}(l,d) = E_{Tx-elec}(l) + E_{amp}(l,d) = \begin{cases} lE_{elec} + l\xi_{fs}d^2 & \text{, if } d < d_0 \\ lE_{elec} + l\varepsilon_{mp}d^4 & \text{, if } d \geq d_0 \end{cases} \tag{4.9}$$

where l is the length of the message in bits, d is the distance between the transmitter and the receiver, d_0 is a threshold value, E_{Tx} is the total energy spent in the transmission of the message, E_{elec} is the radio electronics energy spent per bit in transmitting and receiving a message by an SN, ξ_{fs} and ε_{mp} are energy consumption parameters of the free space and multi-path fading model. Also, the energy required to receive the message is as follows.

$$E_{Rx} = E_{Rx-elec}(l) = lE_{elec}. \tag{4.10}$$

4.3.2 Terminologies and Important Definitions

In this section, we present some terminologies that are used for problem formulation and the proposed work.

Definition 4.3.2.1 Tour of the MS (D_{tour}): A tour is defined as the path traveled by the MS to collect data from the SNs in one hop communication by visiting RPs. It begins and ends with the geographical location of the BS (r_0), i.e.,

$$D_{tour} = \; < r_0, \ldots, r_{n^r}, r_0 > . \tag{4.11}$$

It's length is denoted by $|D_{tour}|$.

Definition 4.3.2.2 Subtour of an MS (D_{stour_i}): A subtour is defined as the path traveled by an MS to collect data from the SNs in one hop communication by visiting a subset of RPs. It begins and ends with the geographical location of the first RP in the subtour (r_1), i.e.,

$$D_{stour_i} = \; < r_1, \ldots, r_t, r_1 > \tag{4.12}$$

The ith subtour length is denoted by $|D_{tour_i}|$. Note also that any two subtours are disjoint, i.e., $D_{stour_i} \cap D_{stour_j} = \phi, 1 \leq i, j \leq v$ and $i \neq j$. Also, t denotes the number of RPs in the ith subtour which varies from subtour to subtour.

Definition 4.3.2.3 Normalized tour length (N_{tour}): It is a modified value of the tour length (D_{tour}) such that the new value becomes comparable with the number of sets in a set cover. The normalized tour length is expressed mathematically as follows.

$$N_{tour} = \left\lceil \frac{1}{|D_{tour}|} \times 10^5 \right\rceil \tag{4.13}$$

The multiplication factor considered above is 10^5 as tour length is also in thousands of meters.

Definition 4.3.2.4 Minimum set cover (*minCov*): It is the minimum number of sets required to cover all SNs

Definition 4.3.2.5 Round: It is the time in which an MS collects data from all the SNs and delivers it to the BS once.

4.4 Problem Formulation

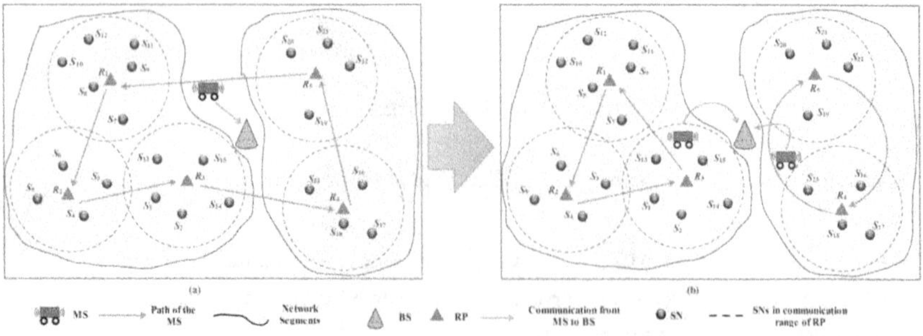

Figure 4.3: Problem depiction:(a) Optimal number of RP selection (b) Optimal number of MS and path determination. (Refer S_i as s_i and R_i as r_i)

The aim of this article is to find optimal number of RPs and their locations coupled with optimal number of MSs and the objective is to reduce the path length as shown in Figs. 4.3(a) and 4.3(b) . The small triangles represent RPs. So, the first task is to find their optimal number as well as locations. Next, the task is find optimal number of MSs such that all RPs are covered and path length is minimum. In the Fig. 4.3, two MSs are shown to cover RPs from the same network segment. However, the RPs traversed by an MS may spread across different network segments also. Since we have considered two tasks, therefore we formulate the problem in two parts as follows.

4.4.1 Problem Formulation for Optimal Number of RPs

Here, we take the comparison of the problem to the min-cost set-covering problem and present the mathematical formulation accordingly. Let x_i and $a_{i,j}$ be the Boolean variables defined as follows:

$$x_i = \begin{cases} 1 & \text{if the set } \phi_i \text{ is included in the cover,} \\ 0 & \text{otherwise} \end{cases} \tag{4.14}$$

and

$$a_{i,j} = \begin{cases} 1 & \text{if the SN } s_i \text{ is covered by the set } \phi_j, \\ 0 & \text{otherwise.} \end{cases} \tag{4.15}$$

Let $b_{i,j}$, O_i, and $h_{i,j}$ be the Boolean variables defined as follows:

$$b_{i,j} = \begin{cases} 1 & \text{if } r_i \text{ and } r_j \in D_{tour}, \\ 0 & \text{otherwise} \end{cases} \tag{4.16}$$

,

$$O_i = \begin{cases} 1 & \text{if } r_i \in D_{tour}, \\ 0 & \text{otherwise} \end{cases} \tag{4.17}$$

$$h_{i,j} = \begin{cases} 1 & \text{if } s_i \text{ is covered in one hop distance by } r_j, \\ 0 & \text{otherwise.} \end{cases} \tag{4.18}$$

Then, the ILP problem is formalized as follows:

$$\text{Minimize } F_1 = w_1 \times minCov + w_2 \times D_{tour} \tag{4.19}$$

where

$$minCov = \sum_{j=1}^{n^r} c_j x_j \tag{4.20}$$

$$D_{tour} = \sum_{1 \leq i,j \leq n^r, i \neq j} d_{i,j} b_{i,j} \tag{4.21}$$

subject to

$$\sum_{j=1}^{n^r} a_{i,j} x_j \geq 1, i = 1, 2, \ldots, n^s \tag{4.22}$$

$$\prod_{r_i \in R} O_i = 1, \tag{4.23}$$

$$\prod_{1 \leq i \leq n^s, 1 \leq j \leq n^r} h_{i,j} = 1, \tag{4.24}$$

$$w_1 + w_2 = 1. \tag{4.25}$$

The constraint in Eq. 4.22 ensures all SNs are covered by the set. The constraint in 4.23 ensures that each of the RPs is included in the tour. It also indicates that each RP is visited only once in the tour. Constraint 4.24 ensures that all SNs are sending data to the RPs in one hop transmission. There can be more than one set cover which provide coverage of all SNs. However, our aim is to find that set which will lead to the shortest path. This set gives an optimal number of RPs ($R = \{r_1, r_2, \ldots, r_{n^r}\}$). The centroid of each cover set acts as the optimal location of an RP.

4.4.2 Problem Formulation for Optimal Number of MSs

Now, we present the second objective to determine the optimal number of MSs such that the total subtour length becomes minimum. Let $b_{i,j}$, and $O_{j,i}$ be the Boolean variables defined as follows:

$$b_{i,j} = \begin{cases} 1 & \text{if } r_i \text{ and } r_j \in D_{stour_t}, \\ 0 & \text{otherwise} \end{cases} \tag{4.26}$$

,

$$O_{j,i} = \begin{cases} 1 & \text{if } r_j \in D_{stour_i}, \\ 0 & \text{otherwise.} \end{cases} \tag{4.27}$$

Then, the ILP problem is formalized as follows:

$$\text{Minimize } F_2 = stdDevStour \tag{4.28}$$

where

$$stdDevStour = \frac{1}{v-1}\sqrt{\sum_{i=1}^{v} \sum_{1 \leq i,j \leq n_i^t, i \neq j} \left(d_{i,j} \times b_{i,j} - \mu\right)^2} \tag{4.29}$$

subject to

$$\prod_{i=1,j=1}^{i=v,j=n_i^t} O_{j,i} = 1, \tag{4.30}$$

$$\sum_{1 \leq i \leq v} R_{stour_i} = R. \tag{4.31}$$

The constraint in Eq. 4.30 ensures that an RP is included in a subtour only once. Constraint 4.31 ensures all RPs are included in the subtours.

4.5 Proposed Work

The basic working steps of the proposed scheme is shown in Fig. 4.4. Each of the steps is described in detail in the subsequent sections. However, let us first start with the set formation process as follows.

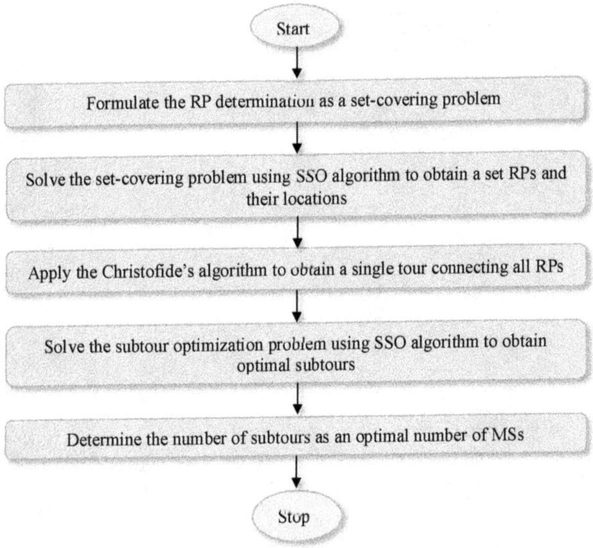

Figure 4.4: Basic working of proposed method.

4.5.1 Set Formation

Let us consider an example of WSN scenario as shown in Fig. 4.5 with 15 SNs, $S = \{s_1, s_2, \dots, s_{15}\}$. The base station is positioned in the center of the deployment area. The SNs are deployed such that some SNs overlap with each other while some are disjointed. The process of set formation proceeds as follows. The first set is considered as all those SNs in the communication range of s_1 (ϕ_1). Then, the second set is considered as the group of SNs in the communication range of s_2 (ϕ_2). This SN is considered for set formation as a new SN (s_4) is in it's communication range. However, this set might not have been considered provided s_4 was absent from here. In a similar manner, s_3 is not considered for set formation as it is already included in ϕ_1 and ϕ_2, and also no new SN is in it's communication range. A new SN indicates that SN which is not yet included in any of the sets so far. Similarly, s_4 and s_5 are considered as new sets in the set formation process; on the other hand, s_6 and s_7 are not considered as sets in the set formation process for the same reason as explained above. Likewise, the other leftover SNs form the remaining sets, and finally, the family of overlapping subsets of S is $\mathscr{C} = \{\phi_1, \phi_2, \dots, \phi_9\}$. The complete process is illustrated in Table 4.1. These sets acts as input to the SSO algorithm and are optimized to obtain the minimal set with shortest path length as explained in the next

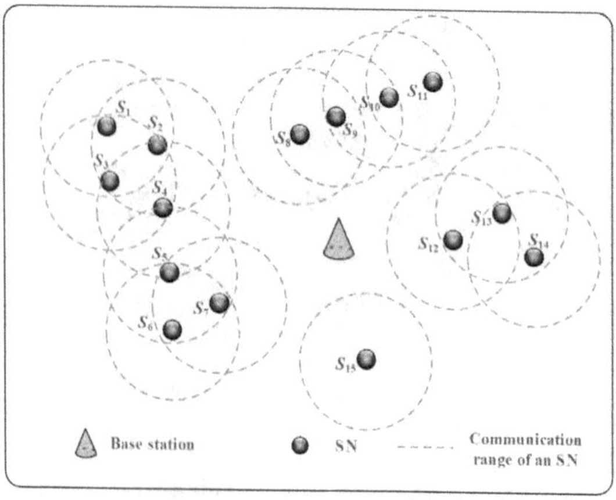

Figure 4.5: A set formation example in a network. (Refer S_i as s_i)

Table 4.1: Set formation process

SN	SNs covered	Considered for set formation	Set number	Remarks
1	1,2,3	Yes	ϕ_1	Considered for set formation
2	1,2,3,4	Yes	ϕ_2	Considered for set formation
3	1,2,3,4	No		Not considered for set formation
4	2,3,4,5	Yes	ϕ_3	Considered for set formation
5	4,5,6,7	Yes	ϕ_4	Considered for set formation
6	5,6,7	No		Not considered for set formation
7	5,6,7	No		Not considered for set formation
8	8,9	Yes	ϕ_5	Considered for set formation
9	8,9,10	Yes		Not considered for set formation
10	9,10,11	Yes	ϕ_6	Considered for set formation
11	10,11	No		Not considered for set formation
12	12,13	Yes	ϕ_7	Considered for set formation
13	12,13,14	Yes	ϕ_8	Considered for set formation
14	13,14	No		Not considered for set formation
15	15	Yes	ϕ_9	Considered for set formation

section.

4.5.2 SSO Based RP Determination

Here, we describe the algorithm based on the SSO algorithm to determine the optimal number of RPs so that all SNs are covered, and the path length is also minimized. The pseudo code for this phase is given in Algorithm 6. In Step 1, we initialize the odour particles and corresponding velocities. In steps 2 and 3, we initialize two variables l_{opt} and g_{opt} to store the local optimum and global optimum values of the population. In

Algorithm 6: Algorithm for SSO-based optimal number of RPs determination

Input : $\mathscr{C}$
Output: $D_{tour}, \mathscr{F}$
1 Initialize the position (X_i^1) and velocity (V_i^1) $(0 \leq x_{i,j}^1, v_{i,j}^1 \leq 1, 1 \leq i \leq n_p, 1 \leq j \leq n_d)$;
2 $l_{opt} \leftarrow$ Find the fitness of each odour particle ;
3 $g_{opt} \leftarrow$ Find the odour particle with maximum fitness i.e., minimum value of l_{opt} ;
4 **for** k = 1 *to* k_{max} **do**
5   Update velocity(V_i^k) using Eq. 4.5 ;
6   Compute shark position (Y_i^{k+1}) using Eq. 4.6;   `// forward movement;`
7   Compute local search particle ($Z_i^{k+1,m'}$) using Eq. 4.7;   `// rotational movement;`
8   Update odour particles (X_i^{k+1}) based on fitness values of Y_i^{k+1} and $Z_i^{k+1,m'}$ using Eq. 4.8 ;
9     $l_{opt} \leftarrow$ Find the fitness values of odour particles (X_i^{k+1}) ;
10   **if** $g_{opt} < minimum(l_{opt})$ **then**
11   update g_{opt} odour particle with the fittest odour particle ;
12   **end**
13 **end**
14 $\mathscr{F} \leftarrow$ Obtain the solution (i.e., set cover) using the g_{opt} odour particle and $\mathscr{C}$;
15 $D_{tour} \leftarrow$ Obtain TSP tour of $\mathscr{F}$ using Christofide's algorithm ;

steps 4 to 13, different parameters of the algorithm (i.e., velocity, shark position (Y), local search particle, odour particles (X), l_{opt}, g_{opt} are updated repeatedly until the termination criteria is reached. In Step 14, the desired set cover ($\mathscr{F}$) is obtained with the help of g_{opt} and $\mathscr{F}$. Finally, in Step 15, the TSP algorithm is applied to $\mathscr{F}$ to obtain a tour of the MS. Note that when the TSP algorithm is applied to $\mathscr{F}$, the in order sequence of sets is designated as the set of RPs.

In the remaining of this section, we describe the first odour particle representation, first illustration, and the first fitness function. However, let us first describe the first odour particle representation.

4.5.2.1 Odour Particle Representation I

We represent an odour particle in such a way that it yields a complete solution to find the optimal number of RPs and their location. Let us consider Fig. 4.6 and Table 4.1. Then, the dimension of the particle is determined as follows. From Table 4.1, s_1 is in sets ϕ_1 and ϕ_2 so for this SN 2 cells in dimension of particle are allocated, s_2 is in sets ϕ_1, ϕ_2, and ϕ_3, so for this SN, 3 cells in dimension of particle are allocated while forming the odour particle. Similarly, for s_3, 3 cells in the dimension of the particle are allocated, for s_4, 3

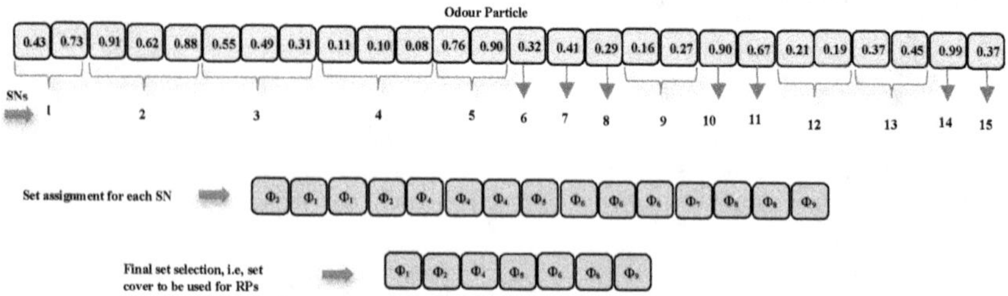

Figure 4.6: Final RPs selection using an odour particle.

cells in the dimension of the particle are allocated, and so on. Finally, the dimension of the particle is $2+3+3+3+2+1+1+1+2+1+1+2+2+1+1=26$. In this way, the particle's dimension is represented, and each cell is given a random real number between 0 and 1, as shown in Fig. 4.6. Next, we demonstrate how a solution is formed using the particle with the help of an illustration.

4.5.2.2 Illustration I

The odour particle is represented, as shown in Fig. 4.6. The solution is obtained from the particle as follows. As two cells are allocated for s_1, so two random real numbers are initialized for the two cells $(0.43, 0.73)$ in the odour particle. s_1 exists in two sets ϕ_1 and ϕ_2, and the 2nd cell of s_1 has the highest value so ϕ_2 is selected. For s_2, 3 cells are allocated and the values initialized are $0.91, 0.63$ and 0.88. The highest among these values is of the 1st cell and since s_2 exists in ϕ_1, ϕ_2, and ϕ_3, so ϕ_1 is selected. Similarly, for all other SNs, the sets are selected. It is worth mentioning that instead of selecting the cell with the highest value and the corresponding set, the cells with the lowest number could also have been selected and processed accordingly. After obtaining all sets corresponding to each of the SNs, there might be some redundant sets. The final solution is obtained by eliminating the redundant sets. The TSP algorithm is then applied to these sets to obtain a sequence such that the path length is minimum. This sequence of the set is mapped as the set of RPs and the centroid of the SNs in a set acts as the location of the corresponding RP.

4.5.2.3 Fitness Function I

We now construct a fitness function to evaluate each of the odour particles as shown in Fig. 4.6. This helps in periodic updation of the local and the global optimal solution. The first objective is to minimize the number of sets needed to cover all SNs, whereas the second objective is to minimize the tour length. Therefore, two objectives are as follows.

$$\text{Objective 1 : Minimize } minCov \tag{4.32}$$

and

$$\text{Objective 2 : Minimize } D_{tour} \tag{4.33}$$

Mathematically, it is written as follows.

$$\text{Minimize } Fitness_1 \tag{4.34}$$

where

$$Fitness_1 = w_1 \times minCov + w_2 \times D_{tour} \tag{4.35}$$

In our approach, $w_1 + w_2 = 1$. The smaller the fitness value of an odour particle, the better is the fitness of the particle. Note that in Eq. 4.34 the value of D_{tour} is very large in comparison to $minCov$, so D_{tour} has been normalized to N_{tour} for the sake of calculation of fitness value.

4.5.3 SSO Based MS Determination

In this section, we describe the algorithm based on the SSO algorithm to determine the optimal number of MSs and also the path of each MS in order to minimize each subtour length such that the latency in data collection is reduced. The pseudo code for this phase is given in Algorithm 7. In Step 1, we initialize the odour particles and corresponding velocities. In steps 2 and 3, we initialize two variables l_{opt} and g_{opt} to store the local optimum and global optimum values of the population. In steps 4 to 13, different parameters of the algorithm (i.e., velocity, shark position (Y), local search particle, odour particles (X), l_{opt}, g_{opt} etc.) are updated repeatedly until the termination criteria is reached. In Step 14, subtours (D_{stour}) are obtained using the g_{opt} of the population which gives the optimal number of MSs and the assignment of RPs to each MS. In Step 15, TSP algorithm is applied to each D_{stour} in order to obtain a tour for each MS. Finally, in Step 16 the number of subtours are determined as the optimal number of MSs (v).

In the remaining of this section, we describe the odour particle representation, an illustration, and the fitness function for Algorithm 7. However, let us first explain the second odour particle representation.

Algorithm 7: Algorithm for SSO-based optimal number of MS determination

Input : $D_{tour}, \mathscr{F}$
Output: D'_{stour}, v

Initialize the position (X_i^1) and velocity (V_i^1) $(0 \leq x_{i,j}^1, v_{i,j}^1 \leq 1, 1 \leq i \leq n_p, 1 \leq j \leq n_d)$;
$l_{opt} \leftarrow$ Find the fitness of each odour particle ;
$g_{opt} \leftarrow$ Find the odour particle with maximum fitness i.e., minimum value of l_{opt};
for k = 1 *to* k_{max} **do**
  Update velocity(V_i^k) using Eq. 4.5 ;
  Compute shark position (Y_i^{k+1}) using Eq. 4.6; `// forward movement;`
  Compute local search particle ($Z_i^{k+1,m'}$) using Eq. 4.7; `// rotational movement;`
  Update odour particles (X_i^{k+1}) based on fitness values of Y_i^{k+1} and $Z_i^{k+1,m'}$ using Eq. 4.8 ;
  $l_{opt} \leftarrow$ Find the fitness values of odour particles (X_i^{k+1}) ;
  if $g_{opt} < minimum(l_{opt})$ **then**
    update g_{opt} odour particle with the fittest odour particle ;
  end
end
$D_{stour} \leftarrow$ Obtain the solution using the g_{opt} odour particle and $\mathscr{F}$;
$D'_{stour} \leftarrow$ Apply TSP algorithm to each subtour in D_{stour} ;
$v \leftarrow |D'_{stour}|$;

4.5.3.1 Odour Particle Representation II

Here, an odour particle is represented such that it yields a complete solution to find the optimal number of MSs and sequence of RPs for each MS. The dimension of the particle is assumed to be the maximum number of MSs to cover the entire network by fulfilling certain constraints such as time limit, path length, etc. We have assumed the constraint to be path length of the MS. The number of MSs is assumed to be the floor value of $\alpha\%$ of the number of RPs. Each dimension of odour particle, in this case, is also initialized with a random real number between 0 and 1. An example of an odour particle is shown in Fig. 4.7.

4.5.3.2 Illustration II

Let us assume that the RP selection algorithm gives 15 RPs, as shown in Fig. 4.7. Let the value of α in this case is 30, so the maximum number of MSs used is 4 as $\lfloor 0.3 \times 15 \rfloor = \lfloor 4.5 \rfloor = 4$. Then, the dimension of the odour particle is 4 as shown in Fig. 4.7. Now, the solution is obtained from this particle as follows. Each dimension of the particle is multiplied by the count of RPs, i.e.,15 and the ceiling value of the number is taken as

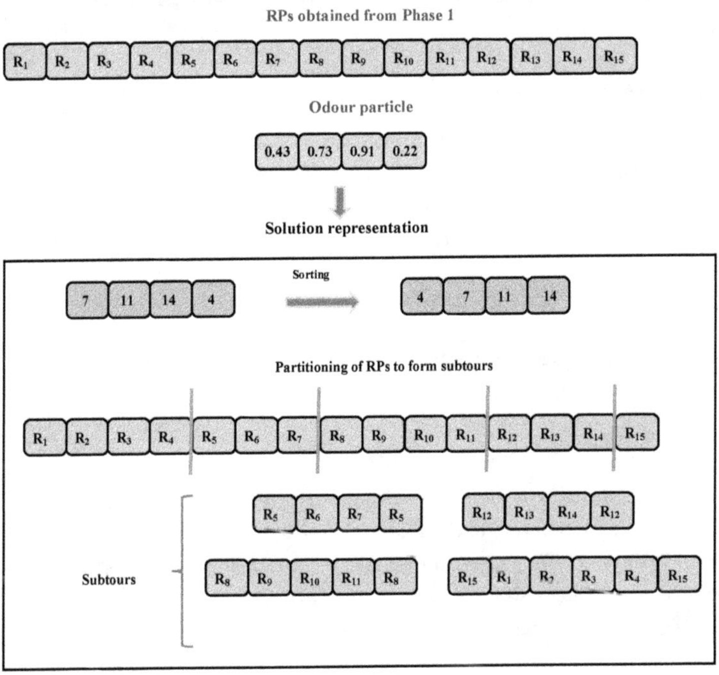

Figure 4.7: Determination of optimal number of MSs and their tours using an odour particle. (Refer R_i as r_i)

follows: $\lceil 0.43 \times 15 \rceil = \lceil 6.45 \rceil = 7$, $\lceil 0.73 \times 15 \rceil = \lceil 10.95 \rceil = 11$, $\lceil 0.91 \times 15 \rceil = \lceil 13.65 \rceil = 14$, and $\lceil 0.22 \times 15 \rceil = \lceil 3.30 \rceil = 4$. It is worth mentioning that instead of *ceiling* value, we can also take *floor* value without having any significant changes. After this step, each dimension of the odour particle has a positive number, which is then arranged in ascending order. Now, each number of the odour particle gives the position of the demarcation of the subtours. Since the first number of the odour particle is 4, and the following number is 7, so the first partition is as follows: $\langle r_5, r_6, r_7 \rangle$. The second number is 7 and the following number is 11, so the second partition is $\langle r_8, r_9, r_{10}, r_{11} \rangle$. Similarly, the next partition is obtained, and for the final partition, the last number is 14, and the following number is 4, so the final partition is $\langle r_{15}, r_1, r_2, r_3, r_4 \rangle$. The starting and ending of the subtours must be the same point, so the first RP in each partition is appended as shown in Fig. 4.7. In this way, the subtours are obtained using the SSO algorithm. There are several combinations of numbers resulting from the odour particle that could lead to subtours with less than three RPs or even one RP; however, we have taken care of all those cases by adequately modifying the process.

4.5.3.3 Fitness Function II

We construct a fitness function to evaluate each of the odour particle as shown in Fig. 4.7. The optimal number of MSs should be obtained such that the overall tour length is minimized and each subtour length is balanced. Also, data collection can be performed with minimum delay. Therefore, the fitness function is written as follows.

$$\text{Minimize: } stdDevStour \tag{4.36}$$

The smaller the fitness value of an odour particle, the better is the fitness of the particle.

4.6 Simulation Results

We performed extensive simulation of our proposed algorithms using Python 3.5 on Ubuntu operating system (64 bit) having Intel(R) Core(TM) i7-8550U CPU 1.80GHz $\times$ 8 processor and 16 GB RAM. We performed simuation by considering two network scenarios, WSN#1 and WSN#2. In WSN#1, the network size was fixed to $1000 \times 1000\ m^2$ and the number of SNs was varied from 200 to 400. While in WSN#2, the number of SNs was fixed to 400 and the network size was varied from $500 \times 500\ m^2$ to $800 \times 800\ m^2$. In both the network scenarios, the communication range was varied to 20 m and 40 m. An SN was assumed to have initial energy of 30 J. The different parameters and their typical values considered for simulation are shown in Table 4.2. We considered two related algorithms, i.e., spanning tree covering algorithm [36] and MCPP algorithm [30] for comparison with our proposed algorithms. Both these algorithms use multiple MS for data collection and try to find the optimal number of MSs. We have used a similar parameter setting, as given in these papers. The parameters used for comparison are the number of RPs, path length, average number of MSs, and energy consumption per round. The topology of the network may vary for different simulation runs; therefore, we have taken an average of 15 different results of various parameters for each of the algorithms to have a fair comparison.

4.6.1 Comparison Based on Number of RPs

In this section, we present the comparison of the proposed algorithm with other existing algorithms in terms of the number of RPs against number of SNs. The rationale for using this parameter is that fewer RPs will ensure lower computational cost as single-hop data

Table 4.2: Simulation parameter setting

Parameter	Value	Unit
Number of SNs (n)	200-600	
Communication range	20,40	m
Network size$(l \times l)$	1000×1000	m^2
Initial energy of SNs	30	J
Packet size	4000	$bits$
E_{elec}	50	nJ/bit
ξ_{fs}	10	$pJ/bit/m^2$
ε_{mp}	0.0013	$pJ/bit/m^4$
E_{DA}	5	nJ/bit
η	0.9	
α	0.1	
β	4	
R'_1	0.5	
R'_2	0.5	
R'_3	0.5	
Number of stages	500	
Population size	100	
Base station location	$(500, 500)$	

collection is performed and shorter path length of MS. Less number of RPs implies the better performance of an algorithm. Fig. 4.8 shows the comparison of the algorithms for WSN#1. It is evident from Fig. 4.8 that our proposed algorithm outperforms the existing algorithms in both cases of communication range, and percentage improvement achieved was 5.69% and 15.64%, respectively over the spanning tree algorithm. It can also be noted that with the increase in the number of SNs, the number of RPs also increase due to their spread over a large area. On comparing the number of RPs in Figs. 4.8(a) and 4.8(b) of the algorithms, it can be observed that the number of RPs in the latter case has decreased significantly for each of the algorithms. This is intuitive because, with an increase in the communication range of an SN, it can communicate data from a long distance, which leads to smaller group formation.

Figs. 4.9(a) and 4.9(b) compare the number of RPs of the three schemes. The communication range was set to 40 m for the simulation scenario of WSN#2. It is evident from the figures that with the increase in network size, the number of RPs also increases. The reason is that with the increase in network size, the spread of SNs also increases, and more RPs are required to cover the SNs. It is noted that the proposed algorithm achieved 5.42% and 12.64% reduction in the number of RPs over the spanning tree and MCPP algorithms, respectively.

Fig.4.10 plots the average number of RPs per MS for WSN#1. This parameter indicates a uniform assignment of RPs to each MS, resulting in an acceptable delay

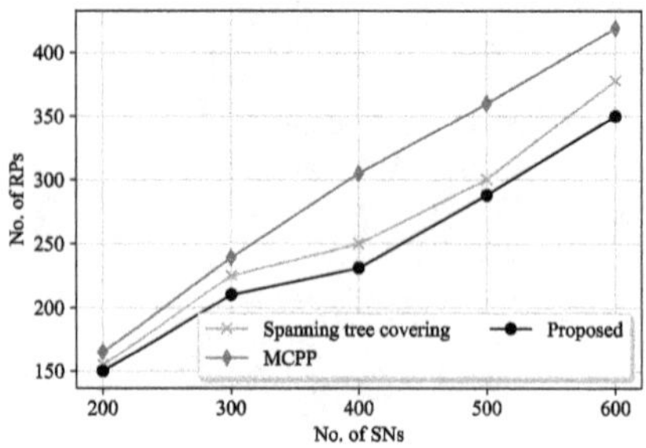

(a)

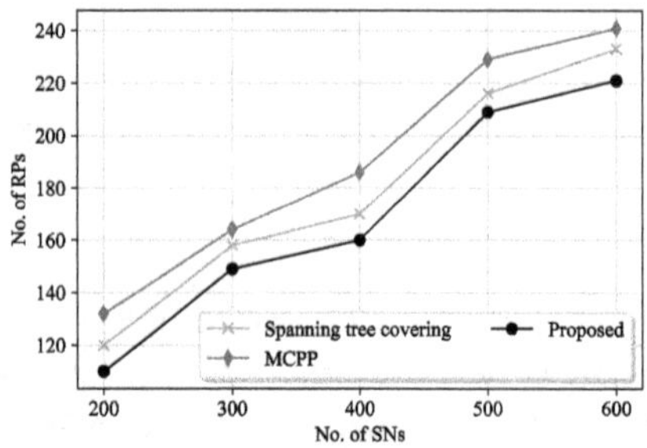

(b)

Figure 4.8: Comparison in terms of number of RPs for (a) $c^s = 20\ m$, (b) $c^s = 40\ m$.

of data delivery from SNs to MS at an RP. A lower value of this parameter suggests that the algorithm is better. It is evident from Fig. 4.10 that our proposed algorithm outperforms the existing algorithms in both cases of communication range, and percentage improvement achieved was 8.61% and 14.89%, respectively over the spanning tree algorithm.

4.6.2 Comparison Based on Path Length

Here, we present the comparison of the proposed algorithm with other existing algorithms in terms of the path length of an MS. The rationale for using this parameter is that a shorter path length will ensure lower computational cost. A shorter path length of an MS implies the better performance of an algorithm. Fig. 4.11 shows the comparison of the algorithms for WSN#1. It is evident from Fig. 4.11 that our proposed algorithm

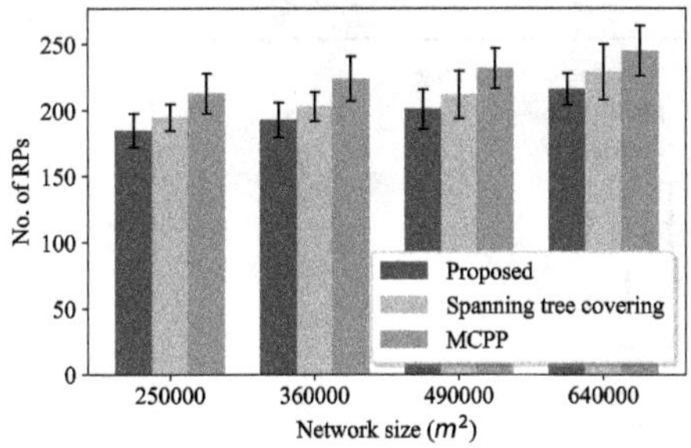

(a)

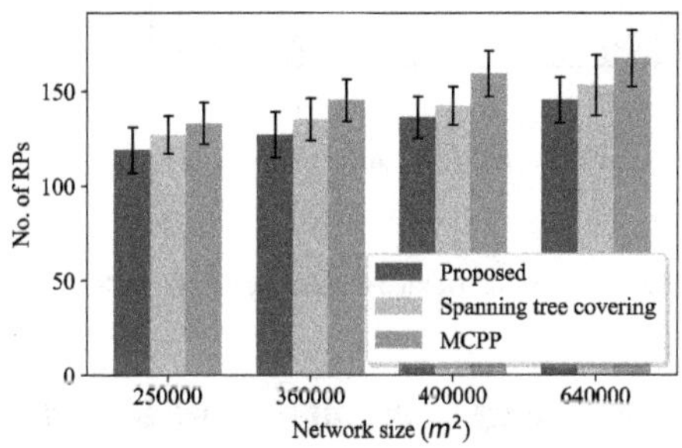

(b)

Figure 4.9: Comparison in terms of number of RPs for (a) $c^s = 20\ m$, (b) $c^s = 40\ m$.

outperforms the existing algorithms for both of the cases of communication range and percentage reduction achieved was 5.53% and 13.87%, respectively over the spanning tree algorithm. From Figs. 4.11(a) and 4.11(b) it can be noted that the path length in the latter case has decreased significantly. This is intuitive because, with an increase in communication range of an SN, it can communicate data even from a long distance.

In Figs. 4.12(a) and 4.12(b), the comparisons of the proposed algorithm with other algorithms in terms of the path length are plotted. Here also, the communication range was set to 40 *m* in the simulation scenario of WSN#2. From the figures, it is clear that with the increase in network size, the path length also increases. The reason is that with the increase in network size, the number of RPs increases, eventually increasing the path length. Note that the proposed algorithm achieved 13.18% and 22.43% reduction in path length over the spanning tree and MCPP algorithms, respectively.

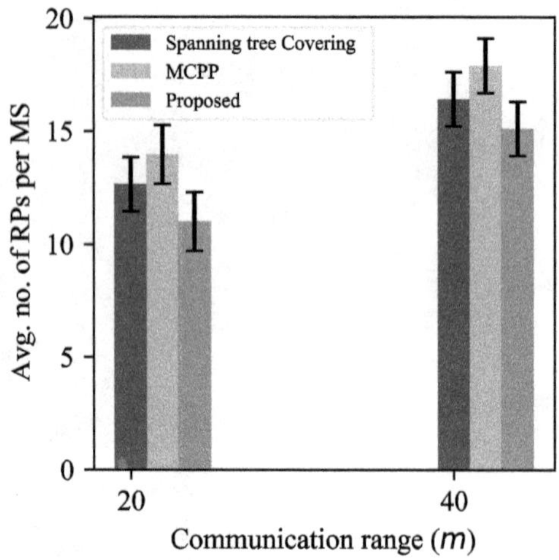

Figure 4.10: Comparison in terms of average number of RPs per MS.

4.6.3 Comparison Based on Energy Consumption per Round

We present the comparison of the proposed algorithm with other existing algorithms in terms of the overall energy consumption per round in this section. This energy expenditure includes the energy spent by MSs while traversing the tour and the energy spent by SNs to communicate data to the MS in one round. The rationale for using this parameter is that less overall energy consumption of SNs would increase the network's lifetime and the efficiency of the system. Lower energy consumption implies the better performance of an algorithm. The energy model used here is explained briefly in section 4.3.1. Fig. 4.13 shows the results of comparison between the algorithms for WSN#1. It is evident from Fig. 4.13 that our proposed algorithm outperforms the existing algorithms in both cases of communication range, and percentage improvement achieved was 19.54% and 29.27%, respectively over the spanning tree algorithm. It can also be noted that with the increase in the number of SNs, the overall energy consumption increase due to an increase in energy expenditure per SN. On comparing the energy consumption per round of MSs in Figs. 4.13(a) and 4.13(b) of the algorithms, it can be observed that the energy consumption per round of MSs in the latter case has increased significantly for each of the algorithms. This is intuitive because more energy per bit is required to send data to a longer distance.

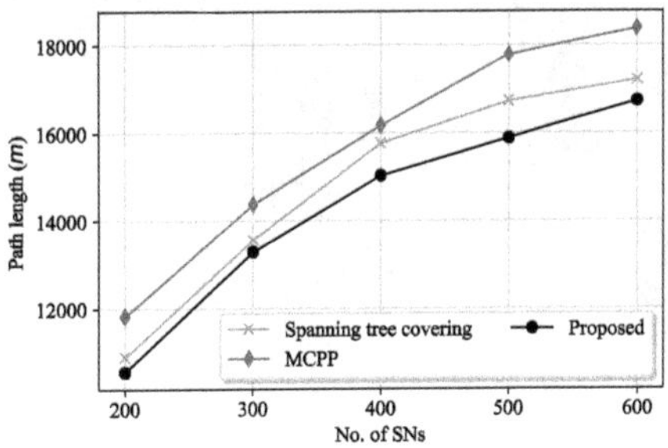

(a)

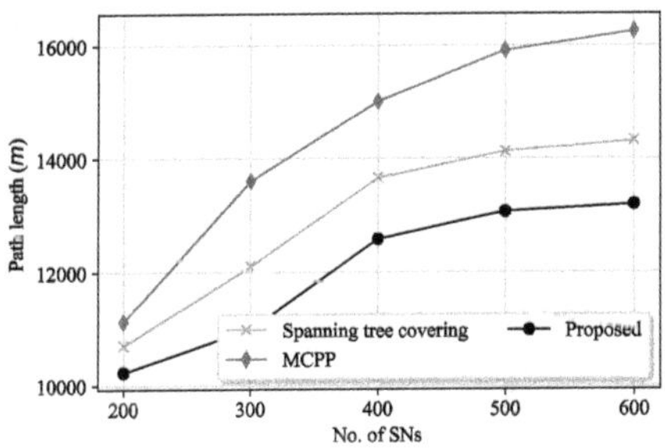

(b)

Figure 4.11: Comparison in terms of path length for (a) $c^s = 20\ m$, (b) $c^s = 40\ m$.

Figs. 4.14(a) and 4.14(b) compare the energy consumption per round of the three approaches in WSN#2 for $c^s = 40\ m$. It can be observed from the figures that with the increase in network size, the energy consumption per round also increases. As explained in Section 4.6.1, the path length increases with the increase in network size; thus, more energy is consumed in traversing the same. Note that the proposed scheme achieved 14.69% and 24.14% less energy per round over the spanning tree and MCPP algorithms, respectively.

4.6.4 Comparison Based on Average MSs

Here, we present the comparison of the proposed algorithm in terms of the average MSs with other existing algorithms. The reason for using this parameter is that fewer MSs would ensure total lower network infrastructure costs. A low number of MSs implies that

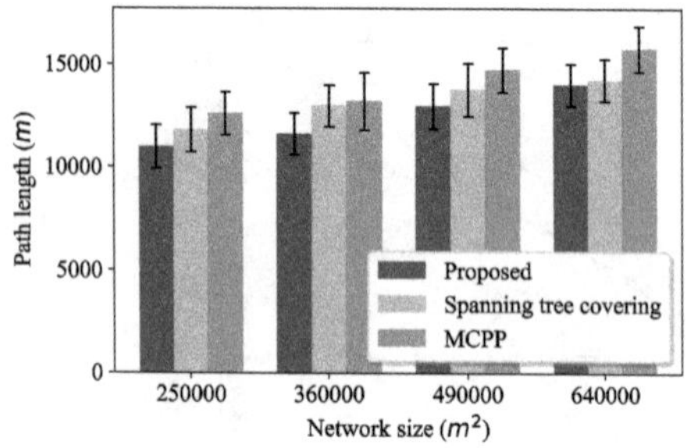

(a)

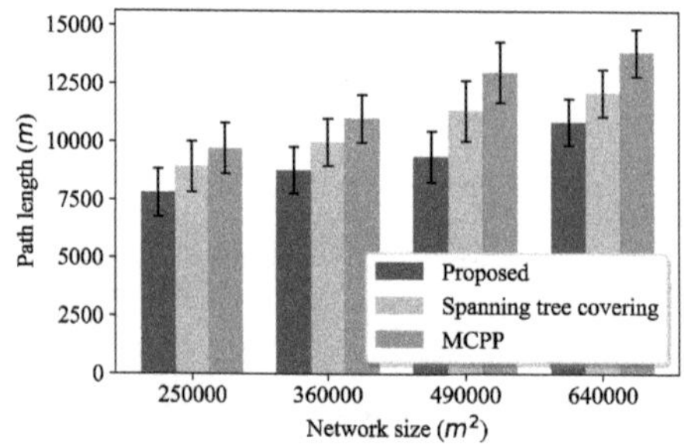

(b)

Figure 4.12: Comparison in terms of path length for (a) $c^s = 20\ m$, (b) $c^s = 40\ m$.

the algorithm is better. Fig. 4.15 shows the comparison of the algorithms for WSN#1. It is evident from Fig. 4.15 that our proposed algorithm outperforms the existing algorithms in both cases of communication range, and percentage improvement achieved was 12.24% and 30.26%, respectively over the spanning tree algorithm. It can also be noted that with the increase in the number of SNs, the number of MSs also increase due to their spread over a large area. On comparing the number of MSs in Figs. 4.15(a) and 4.15(b) of the algorithms, it can also be observed that the number of MSs in the latter case has decreased significantly for each of the algorithms. This is intuitive because the increase in the communication range of an SN leads to smaller group formation, which can be covered by less number of MSs.

Figs. 4.16(a) and 4.16(b) show the comparison of the average number of MSs of the three schemes. It is clear that the average number of MSs increases with the increase in network size. This is because, with the increase in network size, the path length also

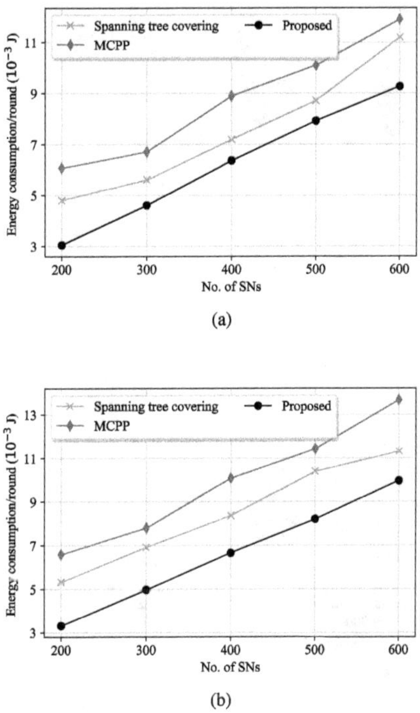

Figure 4.13: Comparison in terms of energy consumption per round for (a) c^s = 20 *m*, (b) c^s = 40 *m*.

increases, leading to more number of MSs for optimal data collection. It is noted that the proposed algorithm achieved 17.81% and 26.46% reduction in the average number of MSs over the spanning tree and MCPP algorithms, respectively.

4.6.5 Convergence Rate of SSO-based Algorithms

Figs. 4.17(a) and 4.17(b) show the convergence rate of the proposed SSO-based Algorithm 6 and Algorithm 7, respectively. For this, the network size was 1000 × 1000 m^2, the number of SN was 600, and the communication range was 40 *m*. The gradual convergence of the plot in Figs. 4.17(a) and Figs. 4.17(b) is attributed to the fact that the exhaustive search of a solution in the form of forward and rotational movement of the shark in SSO leads to reaching the solution relatively faster than other metaheuristic algorithms (e.g., PSO). As a result, the problem of getting trapped in local minima is

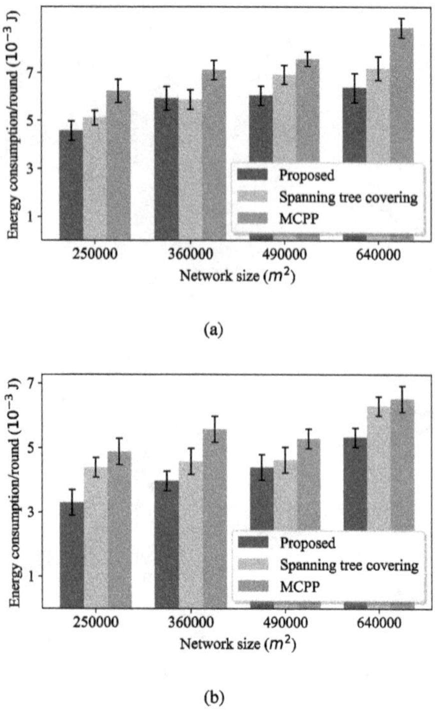

Figure 4.14: Comparison in terms of energy consumption per round for (a) $c^s = 20$ *m*, (b) $c^s = 40$ *m*.

mitigated.

Tables 4.3 and 4.4 show the significance of the proposed scheme in the form of percentage improvement for various metrics with respect to the spanning tree covering and MCPP algorithms. Clearly, the proposed scheme is a superior performer.

4.6.6 Statistical Analysis

Here, we perform one way ANOVA test on the sample data generated using the spanning-tree covering algorithm, MCPP algorithm, and the proposed algorithm (SSO-RP). It is a well known statistical test that determines whether groups of data from two or more schemes are significantly different. To perform the test, we have considered three parameters; the number of RPs, the path length of an MS, and the number of MSs. Here, we have considered the number of SNs to be 600, network size to be

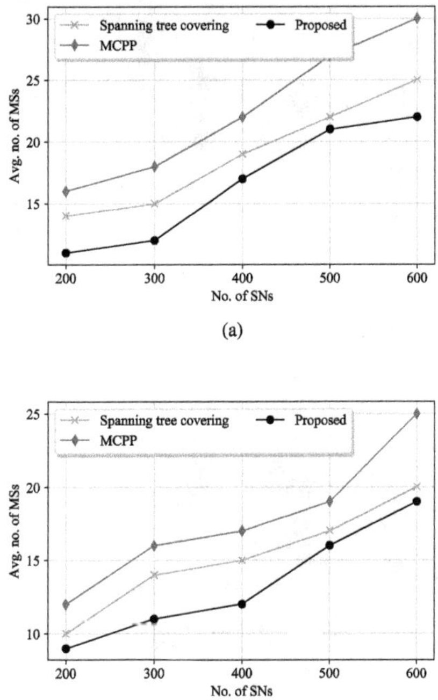

(a)

(b)

Figure 4.15: Comparison in terms of average MSs for (a) $c^s = 20\ m$, (b) $c^s = 40\ m$.

$1000 \times 1000\ m^2$, the communication range of an SN as 40 m, and the sample size to be 15 for each group of data. We consider the null hypothesis as

$$H_0 : \mu_{proposed} = \mu_{spanning\_tree\_covering} = \mu_{MCPP} \tag{4.37}$$

and the alternate hypothesis as

$$H_1 : \mu_{proposed} \neq \mu_{spanning\_tree\_covering} \neq \mu_{MCPP}. \tag{4.38}$$

The null hypothesis (H_0) implies that the means of the group are equal, whereas the alternate hypothesis implies that the means of the group are significantly different. H_0 is rejected if p-value obtained using ANOVA test is less than the α-value (0.05) and F_{stat} is higher than F_α otherwise it is accepted. The results obtained after performing the test are given in Table 4.5. It can be observed that for all three parameters, the above-mentioned

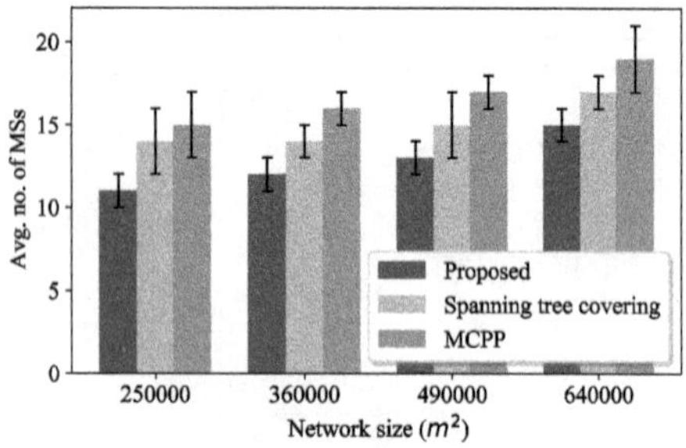

(a)

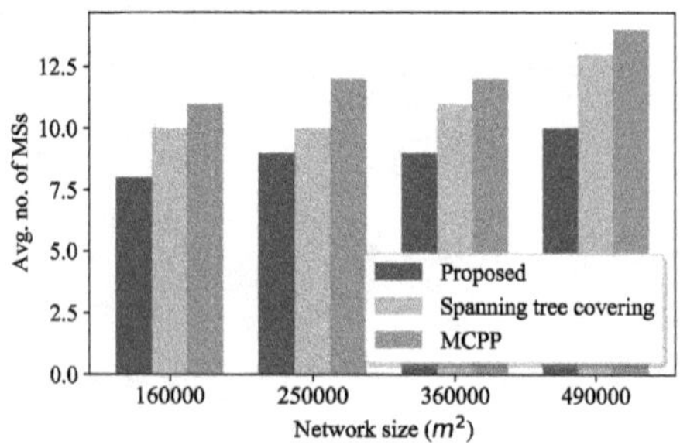

(b)

Figure 4.16: Comparison in terms of average MSs for (a) $c^s = 20\ m$, (b) $c^s = 40\ m$.

conditions are satisfied; hence the null hypothesis is rejected.

With the ANOVA test's help, the significantly different groups cannot be identified, so we perform Tukey's Honest Significant Difference (HSD) post hoc test. The results obtained from performing the test are shown in Table 4.6. The third column of the table shows negative mean difference values for each group of data that are significantly different. Also, the 'Reject' column for each case indicates 'True,' meaning that the null hypothesis is to be rejected. Therefore, it can be clearly concluded that our proposed algorithm is statistically significant and differ from the other two algorithms.

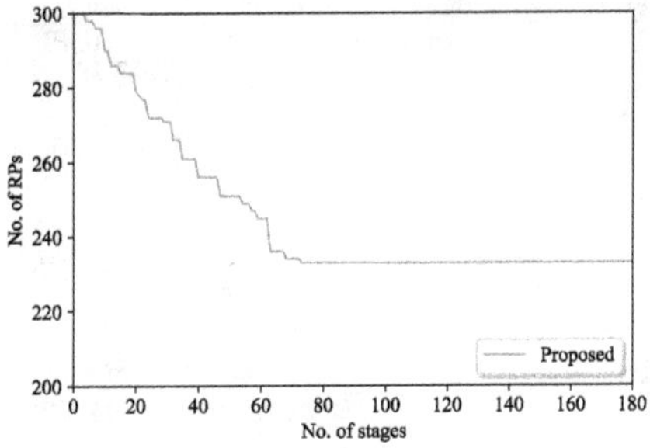

(a)

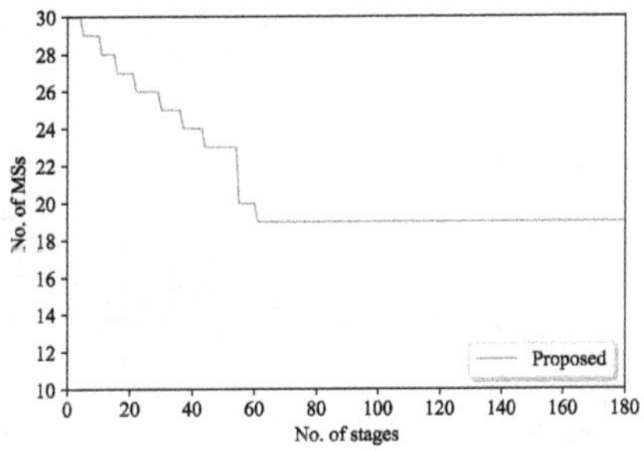

(b)

Figure 4.17: Convergence rate for (a) Algorithm 6 and (b) For Algorithm 7.

Table 4.6: Tukey's HSD post hoc test results

(a) Number of RPs						
Group1	Group2	MeanDiff	p-adj	Lower bound	Upper bound	Reject
MCPP	Proposed	-26.0	0.001	-28.6809	-23.3191	True
Spanning tree covering	Proposed	-17.4	0.001	-20.0809	-14.7191	True
(a) Path length of MS						
Group1	Group2	MeanDiff	p-adj	Lower bound	Upper bound	Reject
MCPP	Proposed	-2857.0667	0.001	-3580.728	-2133.4053	True
Spanning tree covering	Proposed	-1039.1333	0.003	-1762.795	-315.472	True
(a) Number of MSs						
Group1	Group2	MeanDiff	p-adj	Lower bound	Upper bound	Reject
MCPP	Proposed	-4.2	0.001	-5.1562	-3.2438	True
Spanning tree covering	Proposed	-1.6667	0.001	-2.6229	-0.7105	True

Table 4.3: Percentage improvement of the proposed algorithm in WSN#1

Parameter		Spanning tree covering		MCPP	
		Communication range			
		20 m	40 m	20 m	40 m
No. of RPs	No. of SNs	5.78	5.67	16.39	11.36
Path length		3.50	7.37	9.01	16.01
Energy Consumption/round		18.43	23.82	30.66	35.10
Avg. No. of MSs		13.71	12.46	27.24	25.05

Table 4.4: Percentage improvement of the proposed algorithm in WSN#2

Parameter		Spanning tree covering		MCPP	
		Communication range			
		20 m	40 m	20 m	40 m
No. of RPs	Network Size	5.23	5.42	13.04	12.64
Path length		3.50	7.37	9.01	16.01
Energy Consumption/round		8.50	14.69	23.10	24.14
Avg. No. of MSs		15.20	17.81	24.06	26.46

Table 4.5: Results of one-way ANOVA test

(a) Number of RPs						
Source of Variation	Sum of squares	df	Mean of squares	F_{stat}	p-value	F_{α}
Between Groups	5263.60	2	2631.80	288.15	2.97E-25	3.22
Within Groups	383.6	42	9.13			
Total	5647.2	44				
(b) Path length						
Source of Variation	Sum of squares	df	Mean of squares	F_{stat}	p-value	F_{α}
Between Groups	62737548.13	2	31368774.07	47.14	1.84E-11	3.22
Within Groups	27949984.67	42	665475.83			
Total	90687532.8	44				
(c) Number of MSs						
Source of Variation	Sum of squares	df	Mean of squares	F_{stat}	p-value	F_{α}
Between Groups	134.18	2	67.09	57.74	8.84E-13	3.22
Within Groups	48.8	42	1.16			
Total	182.98	44				

4.7 Chapter Summary

This chapter has proposed an SSO-based solution for two problems, one for determining an optimal number of RPs and another for determining an optimal number of MS that jointly ascertains the traveling path of the MSs. The proposed scheme has the following two distinguishing characteristics: first, it is specifically designed for disjoint WSNs, but it is also equally applicable to linked networks; second, it is suitable for delay-tolerant applications where a shorter path length is required. Furthermore, extensive simulations proved the competency of the scheme in terms of the number of RPs, the number of MSs, path length, and energy consumption of SNs. Hypothesis testing and post hoc analysis are also used to validate the results statistically. In the next chapter, we explore the application of MS for data collection in the edge computing paradigm which is a reality nowadays.

Chapter 5

Data Collection via Mobile Edge Computing in Disjoint WSNs

Chapter 5

Data Collection via Mobile Edge Computing in Disjoint WSNs

5.1 Introduction

This chapter proposes a data collection scheme for disjoint networks via mobile edge computing. The scheme uses a realistic propagation model imbibing the log-normal shadowing phenomena. This is noteworthy that using such a communication model will greatly boost the data collection efficiency by reducing signal loss and partial data packet delivery. In addition, it will also improve the accuracy of the system, which is desirable in many delay-harsh applications. The scheme comprises two algorithms. The first one is based on a convex hull approach, whereas the second one is based on a recently developed metaheuristic scheme called Jaya algorithm [65]. The Jaya algorithm is advantageous over other metaheuristic algorithms like PSO [107], [108], because of the following facts: 1) It is simple and computationally efficient, and 2) It does not require any complex set of parameters tuning. We name our proposed scheme as **C**onvex **H**ull and **J**aya **A**lgorithm based **D**ata **C**ollection (CHJADC). In the following, network environment, preliminaries, problem statement, proposed work, and simulation analysis is detailed.

5.2 Network Environment

5.2.1 System Model

The system consists of a three-layer architecture, as shown in Fig. 5.1. The topmost layer is the cloud consisting of multiple servers, the middle layer is the edge server, and the bottom layer includes a set of deployed WSNs. For development of the proposed scheme and for the sake of simplicity, we consider a single WSN consisting of a set of randomly deployed SNs and multiple MSs as shown in Fig. 5.2. Note that this figure is actually the sub-part (bottom layer) of Fig. 5.1. A base station (BS) is placed at the center of the network. It is assumed that the locations of all the SNs are known to the BS using some localization techniques like [109],[110]. The network has disjoint segments indicated by red borders, i.e, the WSN region has isolated groups of SNs (called segments) which are independent from each other. The MSs roam around the disjoint segments and perform data collection at the bottom layer and communicate it to the BS. The BS communicates this data to the edge server which in turn communicates it to a cloud server for further processing. This data flow is shown by arrow in Fig. 5.1. Each MS halts at different RP locations to collect data from SNs with one-hop distance. They have enough storage and energy, and the speed is sufficient to complete the tour in time. The data collection time at each RP is negligible. Note that the traveling path of any MS is not fixed. In other words, a MS can enter any disjoint segment for the sake of data collection. The MSs also perform data aggregation at the RPs, resulting in efficient handling of data in terms of energy and memory space. The energy model used in communication between SNs and MSs is adopted from [102].

5.2.2 An Overview of the Proposed Scheme

Herein, we present an overview of the proposed scheme using a figurative illustration with a WSN having 4 segments as shown in Fig. 5.3. The symbols have the same meaning as in Fig. 5.2. The aim is to find a set of RPs and associated optimal tour of single MS and optimal number MSs with their trajectories. At first, we form one or more convex hulls in each network segment and designate the centroid of each hull as an RP using a Jarvis March [111] based heuristic. This is implemented by Phase 1 of the proposed work as described in Section 5.5.1. This results into 9 RPs as shown in Fig. 5.3(b) which are connected to form trajectory of a single MS as shown in Fig. 5.3(c) (using the same algorithm). This tour is then divided into subtours for optimal number of MSs by applying Jaya metaheuristic based algorithm, as described in phase 2 of Section 5.5.2. This results into subtours of 2 MSs (which are balanced) as shown in Fig. 5.3(d). It is noteworthy

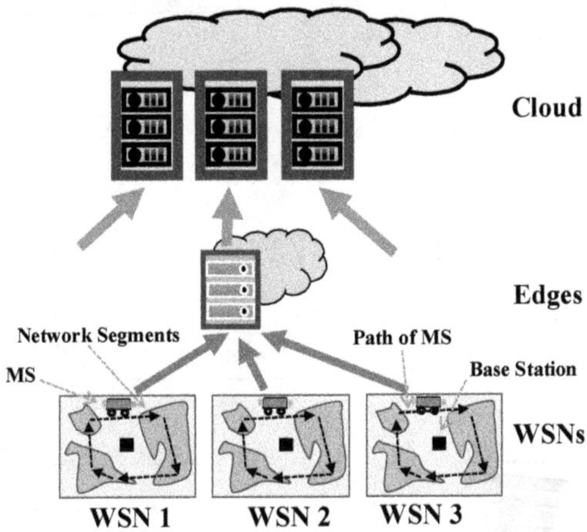

Figure 5.1: A three-layer system architecture: cloud servers, mobile edge, and WSNs equipped with MS.

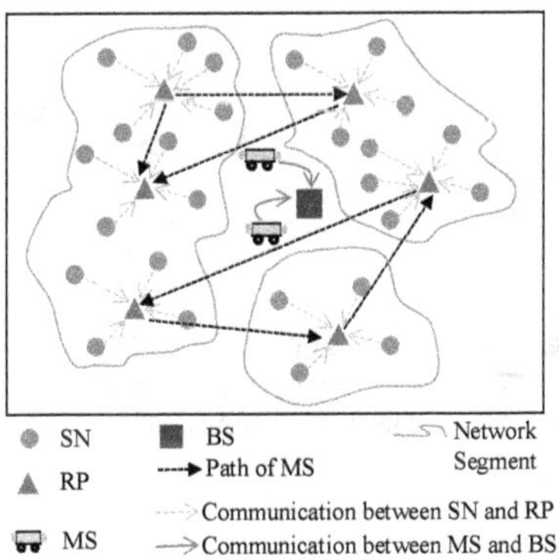

Figure 5.2: Data collection in a disjoint WSN.

that the subtours are spread across different network segments. It is also possible that the subtours may belong to the same network segment.

5.2.3 Propagation Model

Now we describe the propagation model used in our scheme. This is based on log-normal shadowing phenomena which realistically estimates the environment conditions of the

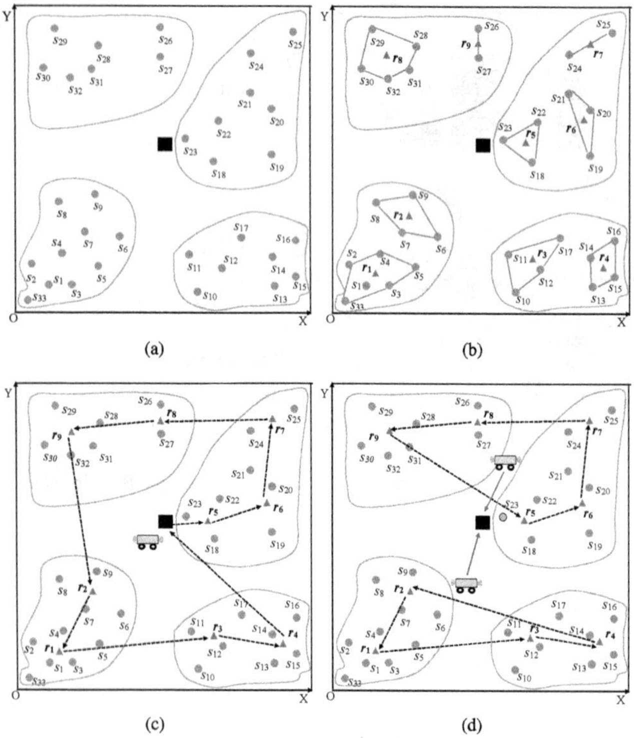

Figure 5.3: Example scenario of the proposed scheme (a) Initial network configuration having multiple MSs and BS, (b) RPs obtained after applying Algorithm 1, (c) Initial tour of single MS using Algorithm 1 , (d) Optimal number of MSs and their balanced tours after applying Algorithm 2.

network for signal propagation from SNs [56]. It is mathematically stated as follows [56]:

$$PL(d) = \overline{PL(d)} + X_\sigma = \overline{PL(d_0)} + 10\eta \log\left(\frac{d}{d_0}\right) + X_\sigma \tag{5.1}$$

and

$$\overline{PR(d)} = PT - \overline{PL(d)} \tag{5.2}$$

where $PL(d)$ is the path loss at a distance d, $\overline{PL(.)}$ is the average path loss, X_σ is the zero-mean Gaussian distributed random variable having standard deviation (σ), d_0 is the close-in reference distance determined empirically close to the transmitter, η is the path loss exponent, $PR(d)$ is the power received at distance d, PT is the power of the transmitter, $\overline{PR(.)}$ is the average power received at distance d.

Next, to determine the probability that a received signal will exceed or fall below a particular threshold value, Q-function or error function (erf) may be used. The definition of Q-function showing it's relation to erf function is as follows [56]:

$$Q(z) = \frac{1}{\sqrt{2\pi}} \int_z^{\infty} exp\left(-\frac{x^2}{2}\right) dx = \frac{1}{2}\left[1 - erf\left(\frac{z}{\sqrt{2}}\right)\right] \tag{5.3}$$

Next, the probability (P_r) that the power of the received signal which exceeds a certain threshold value (γ) can be calculated as follows [56] :

$$P_r(PR > \gamma) = Q\left(\frac{\gamma - \overline{PR(d)}}{\sigma}\right) \tag{5.4}$$

Note that this is related with Boundary coverage which is defined as the circular coverage region with an SN at the center where the mean value of signal strength at a mobile element remains above a threshold value with certain probability.
Similarly, the probability (P_r) that the power of the received signal which falls below a certain threshold value (γ) can be calculated as follows [56]:

$$P_r(PR < \gamma) = Q\left(\frac{\overline{PR(d)} - \gamma}{\sigma}\right) \tag{5.5}$$

5.3 Preliminaries

This section presents some preliminaries required for problem formulation and the algorithms.
Theorem 1: If $\eta = 2$ and $\sigma = 8\ dB$, then for an SN the boundary coverage is 75% and the area coverage is 91%.

Proof. Let R be the radius of the SN and γ be the desired received signal threshold value. Further, let $U(\gamma)$ be the ratio of the useful service area (i.e., the region in which the received signal strength is greater than or equal to γ) to the total area πR^2, i.e., $U(\gamma)$ is a dimension-less quantity. Also, let $d = r$ be the radial distance from the transmitter. Assuming an incremental area dA at distance r, $U(\gamma)$ can be obtained as follows [112]:

$$\begin{aligned} U(\gamma) &= \frac{1}{\pi R^2} \int P_r(PR > \gamma) dA \\ &= \frac{1}{\pi R^2} \int_0^{2\pi} \int_0^R P_r(PR > \gamma) r dr d\theta \end{aligned} \tag{5.6}$$

This equation relates the area coverage which is determined by the radius R at which the signal strength exceeds the threshold value γ.
The path loss with reference to the cell boundary ($r = R$) is [112]

$$\begin{aligned}\overline{PL(r)} =& \overline{PL(d_0)} + 10\eta \log\left(\frac{r}{d_0}\right) \\ =& \overline{PL(d_0)} + 10\eta \log\left(\frac{r}{R}\right) + 10\eta \log\left(\frac{R}{d_0}\right)\end{aligned} \tag{5.7}$$

Since radial distance r is used instead of d, therefore Eq. 5.5 is rewritten as follows :

$$\begin{aligned}P_r(PR > \gamma) &= Q\left(\frac{\gamma - \overline{PR(r)}}{\sigma}\right) = \frac{1}{2} - \frac{1}{2}erf\left(\frac{\gamma - \overline{PR(r)}}{\sigma\sqrt{2}}\right) \\ &= \frac{1}{2} - \frac{1}{2}erf\left(\frac{\gamma - [PT - (\overline{PL(d_0)} + 10\eta\log(r/d_0))]}{\sigma\sqrt{2}}\right) \\ &= \frac{1}{2} - \frac{1}{2}erf\left(\frac{\gamma - [PT - (\overline{PL(d_0)} + 10\eta\log(\frac{R}{d_0}) + 10\eta\log(\frac{r}{R}))]}{\sigma\sqrt{2}}\right) \\ &= \frac{1}{2} - \frac{1}{2}erf\left(\frac{\gamma - PT + \overline{PL(d_0)} + 10\eta\log(\frac{R}{d_0}) + 10\eta\log(\frac{r}{R})}{\sigma\sqrt{2}}\right)\end{aligned} \tag{5.8}$$

Assuming $a = (\gamma - PT + \overline{PL(d_0)} + 10\eta\log(\frac{R}{d_0}))/\sigma\sqrt{2}$ and $b = (10\eta\log e)/\sigma\sqrt{2}$, $U(\gamma)$ can be rewritten as [112]

$$U(\gamma) = \frac{1}{2} - \frac{1}{R^2}\int_0^R r\, erf\left(a + b\ln\frac{r}{R}\right)dr \tag{5.9}$$

On substituting variable $t = a + b\ln(r/R)$ in Eq. 5.9 and solving we get

$$U(\gamma) = \frac{1}{2}\left(1 - erf(a) + exp\left(\frac{1-2ab}{b^2}\right)\left[1 - erf\left(\frac{1-ab}{b}\right)\right]\right) \tag{5.10}$$

Now, choosing the signal level such that $\overline{PR(R)} = \gamma$ (i.e., $a = 0$) $U(\gamma)$ can be written as follows:

$$U(\gamma) = \frac{1}{2}\left[1 + exp\left(\frac{1}{b^2}\right)\left(1 - erf\left(\frac{1}{b}\right)\right)\right] \tag{5.11}$$

Finally, substituting the values of $\eta = 2$ and $\sigma = 8dB$ to obtain corresponding value for b in Eq. 5.11, the boundary coverage turns out to be 75% whereas area coverage comes out to be 91%. □

This theorem helps to set the communication range of the SNs to achieve the desired signal strength for transmitting sensed data to the RPs.

Definition 5.3.1 (*2-argument arctangent*): The 2-argument arctangent function also known as *arctan*2 function is defined as follows:

$$arctan2(y,x) = \begin{cases} arctan(\frac{y}{x}) & \text{if } x > 0, \\ arctan(\frac{y}{x}) + \pi & \text{if } x < 0 \text{ and } y \geq 0, \\ arctan(\frac{y}{x}) - \pi & \text{if } x < 0 \text{ and } y < 0 \end{cases} \tag{5.12}$$

where $x = r\cos\theta$ and $y = r\sin\theta$, $r = \sqrt{x^2+y^2}$, and θ is the angle between the two points as shown in Fig. 5.4.

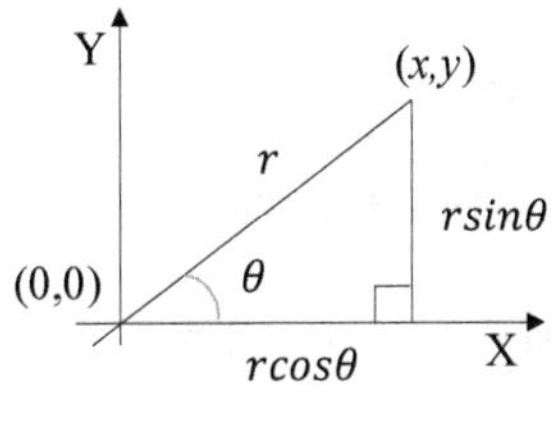

Figure 5.4: The angle theta.

Definition 5.3.2 (*Tour of MS*): A tour of a MS is known as the route it takes to travel for data collection through the deployment region by moving from RP to the other RP. The starting and ending location of the MS is BS (i.e., r_0). It is expressed mathematically as follows:

$$D_{tour} = < r_0, r_1, r_2, \ldots, r_m, r_0 > \tag{5.13}$$

and the length of the tour is

$$|D_{tour}| = \sum_{i=1}^{m-1} d_{i,i+1} + d_{0,1} + d_{m,0} \tag{5.14}$$

where $m = n^r$ = total number of RPs.

Definition 5.3.3 (*Subtour of MS*): A subtour of a MS, denoted by D_{stour_i} is defined as a subset of RPs in D_{tour} . It begins and ends with the location of the first RP in the subtour. The summation of all subtour lengths of all MSs is defined as

$$\begin{aligned} |D_{stour}| &= \sum_{i=1}^{v} |D_{stour_i}| \\ &= \sum_{i=1}^{v} \sum_{j=1}^{n_i^t} d_{j,j+1} + d_{r_l^i, r_1^i} \end{aligned} \tag{5.15}$$

5.4 Problem Formulation

Given a set of disjoint network segments, we are supposed to find a set of RPs and an optimal tour of a single MS along with an optimal number of MSs with their trajectories. As discussed earlier, we first derive a set RPs to form an optimal tour of a single MS. Then the tour is divided into subtours for an optimal number of MSs. Note that the subtour for each MSs must also be optimal. We now present problem formulations for the same successively as follows.

5.4.1 Optimizing Tour of a Single MS

Let $b_{i,j}$, O_i, and $h_{i,j}$ be the Boolean variables defined as follows:

$$b_{i,j} = \begin{cases} 1 & \text{if } r_i \text{ appears immediately before } r_j \text{ in } D_{tour}, \\ 0 & \text{otherwise.} \end{cases} \tag{5.16}$$

$$O_i = \begin{cases} 1 & \text{if } r_i \text{ appears only once in } D_{tour}, \\ 0 & \text{otherwise.} \end{cases} \tag{5.17}$$

$$h_{i,j} = \begin{cases} 1 & \text{if } s_i \text{ is assigned to only } r_j, \\ 0 & \text{otherwise.} \end{cases} \tag{5.18}$$

Our objective is to minimize the tour length of the MS so that overall energy consumption and latency can be reduced, which in turn enhance the network lifetime. Therefore, we have the following optimization problem:

$$\text{Minimize } |D_{tour}| \ (refer\ to\ Eq.\ 5.14) \tag{5.19}$$

subject to

$$\prod_{r_i \in R} O_i = 1 \tag{5.20}$$

$$\bigcup_{1 \leq i \leq n^r, 1 \leq j \leq n^{seg}} S^{i,j} = S \tag{5.21}$$

where $S^{i,j}$ denotes the set of SNs covered by r_i of the jth network segment.

$$\sum_{1 \leq i \leq n, 1 \leq j \leq n^r} h_{i,j} = 1. \tag{5.22}$$

where

$$|D_{tour}| = \sum_{1 \leq i,j \leq n^r, i \neq j} d_{i,j} b_{i,j} \tag{5.23}$$

The constraint (5.20) ensures that each RP must appear in the tour only once except the BS (i.e., r_0). The constraint (5.21) ensures that all SNs are covered by the RPs in all of the network segments. The constraint (5.22) guarantees that an SN can send data to a single RP only in the one-hop distance. This also ensures that all the SNs will send their data to their corresponding RPs. Next, we describe the subtour optimization problem for optimizing the number of MSs.

5.4.2 Optimizing Number of MSs

Let $b'_{i,j}$, and $O'_{i,j}$ be the Boolean variables defined as follows:

$$b'_{i,j} = \begin{cases} 1 & \text{if } r_i \text{ appears immediately before } r_j \text{ in } D_{stour_t}, \\ 0 & \text{otherwise.} \end{cases} \tag{5.24}$$

$$O'_{i,j} = \begin{cases} 1 & \text{if } r_i \text{ appears only once in } D_{stour_j}, \\ 0 & \text{otherwise.} \end{cases} \tag{5.25}$$

Also, let Γ be the standard deviation of the lengths of all subtours given by

$$\Gamma = \sqrt{\frac{1}{v-1} \sum_{i=1}^{v} \sum_{1 \leq i,j \leq n_i^r, i \neq j} \left(d_{i,j} \times b'_{i,j} - \mu \right)^2} \tag{5.26}$$

and Λ be the number of MSs. Here, v refers to the number of MSs.

Here, our objective is to find out an optimal number of MSs, and at the same time, the subtour lengths are also to be balanced. To achieve this, we need to minimize the variation in all subtour lengths and also to minimize the number of MSs. Thus, minimizing their linear combination will serve the purpose. Therefore, we have the following problem formulation:

$$\text{Minimize } w_1 \Gamma + w_2 \Lambda \tag{5.27}$$

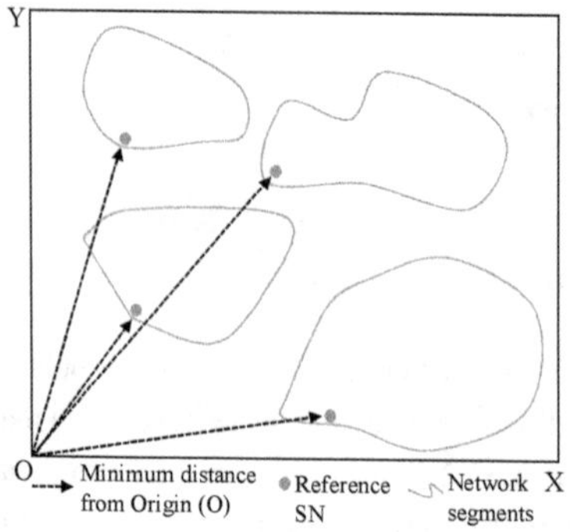

Figure 5.5: The first reference SN (s^r) of each network segment.

subject to

$$\prod_{1 \leq i \leq v, 1 \leq j \leq n_i^t} O'_{i,j} = 1, \tag{5.28}$$

and

$$w_1 + w_2 = 1 \tag{5.29}$$

The constraint in Eq. 5.28 ensures that an RP is included in a subtour only once except the starting RP of each of the subtours. w_1 and w_2 are the weights associated with Γ and Λ whose value is chosen from [0, 1]. Note that an optimal number of subtours will be the optimal number of MSs.

5.5 Proposed Work

The scheme consists of the following two phases 1) convex hull-based optimal RP determination, 2) Jaya algorithm-based optimal number of MS determination. These phases are successively described as follows.

5.5.1 Phase 1: Determining RPs and Optimal Tour of a Single MS Based on Convex Hull

The basic idea here is to ascertain a set of RPs for SNs by iteratively dividing each of the network segments into multiple convex hulls. However, at first a Breadth-First Search (BFS) [113] algorithm determines the number of isolated network segments based on the locations of the SNs. Note that BFS is a well-known algorithm to find out connected components that can be employed to find out the number of isolated segments when their number is not known apriori [113]. After that, in each network segment, a reference SN (s^r) is used to form the first convex hull. For this, an SN located at the edge of this segment is required. Therefore, an imaginary axis is created at the lower left part of the deployment region. The lower-left corner is considered as the origin (O). The horizontal line of the deployment region is considered as the X-axis (OX), whereas the vertical line of the same is considered as the Y-axis (OY). Now, for a network segment (δ_i), the SN located at the edge of this segment is determined as follows:

$$s^r = \{s_j | min(\sqrt{(x_j)^2 + (y_j)^2}, \forall s_j \in S_i^s, \\ 1 \leq j \leq |S_i^s|\} \tag{5.30}$$

where the co-ordinates of s_j are (x_j, y_j) and the co-ordinates of O are $(0,0)$. Note that if for δ_i, s^r has more than one SN, then anyone can be used for the purpose. However, this situation is highly unlikely. In this way, the first reference SN (s^r) of each network segment is determined (see Fig. 5.5).

After that, for δ_i (for the ith network segment), all SNs which are utmost $\alpha \times c^s$ (explained later) distance apart from s^r are determined as follows:

$$S_i^h = \{s_j | \sqrt{(x_i - x_j)^2 + (y_i - y_j)^2} \leq \alpha \times c^s, \\ \forall s_j \in S_i^s \; and \; i \neq j\} \tag{5.31}$$

where (x_i, y_i) are the X and Y co-ordinates of SN in s^r. A convex hull is built with the SNs in $S_i^h + s^r$ using the Jarvis's march algorithm [111] as:

Step 1: Select an origin point as s^r.

Step 2: Find the point (s_i') with the lowest polar angle (θ_i) in the right side following anticlockwise direction until the highest vertical point is reached after that find the point with the lowest polar angle (θ_i) in left side in anticlockwise direction. The lowest polar angle is determined as:

$$\theta_i = min\{arctan2(\frac{y_j}{x_j}), j \in 1, 2, \ldots, n_{i,j}^h \text{ and } i \neq j\} \tag{5.32}$$

where $n_{i,j}^h$ is determined from $S_i^h + s^r$.

Step 3: Shift the origin to the point s_i' and repeat Step 2 until all points in S_i^h (i.e., SNs) are enclosed by the convex hull.

This way the first convex hull is formed and the centroid of coordinates of SNs in $S_i^h + s^r$ are used for the location of the first RP as follows:

$$r_1 = \left(\frac{\sum_{i=1}^{n^h_{1,1}} x_i}{n^h_{1,1}}, \frac{\sum_{i=1}^{n^h_{1,1}} y_i}{n^h_{1,1}} \right) \tag{5.33}$$

Note that the term $\alpha \times c^s$ in Eq. 5.31 ensures communication between SNs and MSs at an RP happens seamlessly considering the log-normal shadowing phenomena. The threshold value α may vary depending on the hardware capability of the SNs. However, we set the value of α to 0.7 based on Theorem 1.

The next convex hull in δ_i is formed with the help of a new reference SN (s'^r). To determine this, two SNs are needed: (a) the farthest SN to s^r in the last formed convex hull located either in the positive X-direction or positive Y-direction (say s^f), and (b) nearest unprocessed (i.e., not enclosed by any convex-hull so far) SN to s^f (say s^n). In order to find s^f, the SNs in the last convex hull are projected onto a new imaginary axis with s^r as origin (O') and parallel to OX (i.e., $O'X'$) or OY (i.e., $O'Y'$). The decision about projection onto $O'X'$ or $O'Y'$ is taken based on the availability of an unprocessed SN node in that direction. However, if unprocessed SNs are available in both directions, then either can be chosen. Note that taking projections and finding distances reduces the computational overhead. Let us choose the positive X-direction, i.e., along the $O'X'$ axis. Then, the desired SN on $O'X'$ axis is mathematically found as follows:

$$\begin{aligned} s^f =& max\left\{ d^s_{i,j} cos\left(arctan2(\frac{(x_j - x_i)}{y_j - y_i} \times (\frac{180}{\pi}))\right\}, \\ & 1 \leq j \leq n^h_{1,1} \ and \ i \neq j \end{aligned} \tag{5.34}$$

where (x_i, y_i) are the coordinates of s^r. The jth point of the convex hull is the desired point along the $O'X'$ axis. The nearest SN to this node is found as:

$$\begin{aligned} s^n =& min\left\{ d^s_{i,j}, i = index(s^f) \ and \right. \\ & \left. j \in \{1, 2, \ldots, |\delta_1|\} \right\} \end{aligned} \tag{5.35}$$

where $index(s^f)$ denotes the SN to start formation of the next convex hull (i.e., $s'^r = s^n$), and $|\delta_1|$ indicates the total number of SNs in δ_1. Note that in the above equation Eq. 5.35, for j those SNs which are already included in any convex-hull of the network segment are excluded. Then, the coordinates of SNs in S^h_i is determined as in Eq. 5.31. Finally, the centroid of the co-ordinates of SNs in $S^h_i + s'^r$ are found and designated as the next RP. The formation of the next convex hull starts with the exchange of coordinates between the previous reference SNs $s'^r \leftarrow s^r$ using the process explained above.

Algorithm 8: Algorithm to determine RPs and an optimal tour of single MS

Input : S, Φ
Output : R, D_{tour}
1 $\Phi \leftarrow$ connected components of S using Breadth First Search algorithm
2 $n^{seg} \leftarrow$ Number of segments in Φ
3 **for** $i = 1$ *to* n^{seg} **do**
4     $s^r \leftarrow$ the reference SN in δ_i
5   **while** *all SNs in* δ_i *are processed* **do**
6         $s'^r \leftarrow s^r$
7         $S_i^h \leftarrow$ all SNs which are $\alpha \times c^s$ distance from s'^r
8   Form convex hull with SNs in $S_i^h + s'^r$ using Jarvis' march algorithm
9         $r^p \leftarrow$ Find the centroid of all SNs in the convex hull previously formed
10   Mark all SNs in the convex hull as processed
11         $R \leftarrow R \cup r^p$
12         $s^r \leftarrow$ Find the next reference SN
13   **end**
14 **end**
15 Return R
16 $D_{tour} \leftarrow$Find approximate shortest tour using R by applying algorithm in [36]

The above-mentioned process of forming convex hulls is repeated for all SNs going from one segment to the other until all the SNs are processed. The pseudo-code of the proposed RP determination method is given in Algorithm 8.

Explanation and time complexity of Algorithm 8: Line 1 uses BFS algorithm with time $O(m' + n^s)$ where m' is the number of wireless links between the SNs and n^s is the number of SNs. Line 3 is to find the least distant SN (using Euclidean distance) from the origin O for a network segment, therefore it will take cost $O(n^{ns})$ time as linear search method is involved. Inside the while loop, line 5 is assignment operation and thus takes $O(1)$ time. Assuming, $n^{ns} << n^s$, $(n^{ns})^2$ comparisons are required to find the SNs for building convex hull (Line 6). Line 7 requires $O(n^{ch}h)$ time to form the convex hull by involving the Jarvis's march algorithm, where h is the number of vertices in the convex hull, and n^{ch} is the average number of SNs in a convex hull($n^{ch} < n^{ns}$). Lines 8, 9, and 10 require $O(1)$ time, respectively. Similar to Line 3, Line 11 requires $O(n^{ns})$ time. Summing the time complexity of while loop we get $O(1) + O((n^{ns})^2) + O(n^{ch}h) + 3 \times O(1) + O(n^{ns}) \approx O((n^{ns})^2) + O(n^{ch}h)$. Let us assume that the number of network segments be n^{seg}. Therefore, while loop is iterated n^{seg} times and finding the approximate shortest tour takes $O((n^r)^2)$ time [36]. Then, the overall time complexity of Algorithm 8 is as follows: $O(m' + n^s)O(n^{seg}) + O(n^{seg}(n^{ns})^2) + O(n^{seg}n^{ch}h) + O((n^r)^2) \approx O(m' + n^s) + O((n^{ns})^2) + O(n^{ch}h) + O((n^r)^2)$.

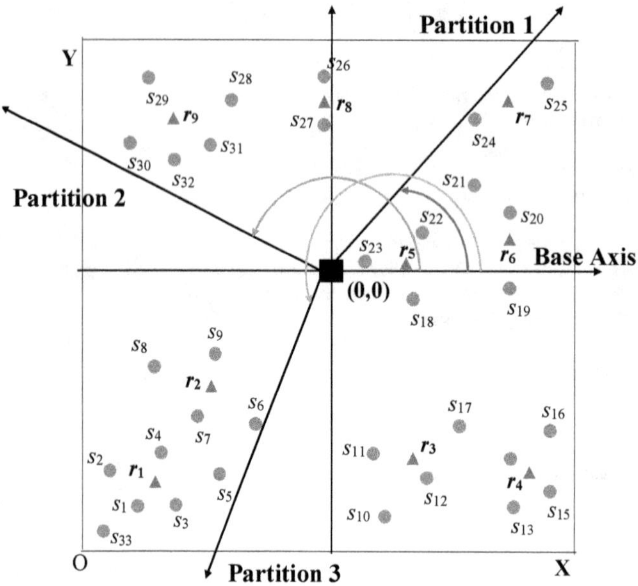

Figure 5.6: Partitioning scheme of RPs with respect to angles formed using an example particle.

5.5.2 Phase 2: Determining Optimal Number of MSs Using Jaya algorithm

Jaya algorithm [65] is a simple and powerful technique, which creates many solutions called population, iteratively refines them and the best solution is reported as the optimal solution. The population of the algorithm is a collection of n_p solutions. The kth solution in j^{th} dimension of the ith iteration is denoted as $X_{j,k,i}$. This value is modified with the help of the following equation:

$$X'_{j,k,i} = X_{j,k,i} + r'_{1\,j,i}(X_{j,best,i} - |X_{j,k,i}|) \\ - r'_{2\,j,i}(X_{j,worst,i} - |X_{j,k,i}|) \tag{5.36}$$

where $X_{j,best,i}$ is the value of the best candidate in j^{th} dimension for the ith iteration, $X_{j,worst,i}$ is the value of the worst candidate in j^{th} dimension for the ith iteration, $r'_{1\,j,i}$ and $r'_{2\,j,i}$ are two random numbers in the range $[0,1]$.

The subtour optimization problem has many possible solutions. However, with the increase in the number of RPs, the potential solutions increases exponentially, contributing to the hardness of the problem. In such cases, meta-heuristic algorithms

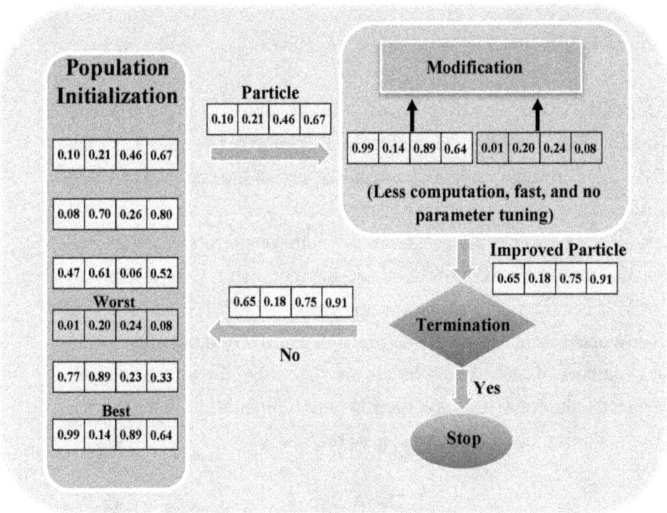

Figure 5.7: Numerical illustration of the Jaya algorithm.

have proven to provide approximately good solutions. Therefore, the Jaya algorithm has been used to achieve an optimal solution for the same.

Here, we use the RPs obtained by Algorithm 8 as input for finding out the optimal number of MSs. We assume that the maximum number of MSs is certain percentage (say 30%) of the number of RPs, i.e., n^r and this can be expressed as $\lceil \beta \times n^r \rceil$ where $\beta = 0.3$. As shown in Fig. 5.3(b), there are 9 RPs, so the dimension of the particle is $\lceil 0.3 \times 9 \rceil = 3$. Now, as a particle in the Jaya algorithm represents a complete solution so for particle representation, each component $(X_{j,k,i})$ of the particle is initialized with a random real number between 0 and 1. For example a particle may be expressed as $[0.43, 0.89, 0.01]$. To obtain a solution from a particle, each component of the particle is multiplied with 360°, and the resulting value is taken at an angle from the base axis as a partition, as shown in Fig. 5.6. The angles using the example particle $[0.43, 0.89, 0.01]$ are $0.43 * 360 = 154.8°$, $0.89 * 360 = 320.4°$, and $0.01 * 360 = 3.6°$. Now, a line is drawn from the origin at each of these angles to create partitions 1, 2, and 3. We start from partition 1 in an anticlockwise direction until partition 2 is reached; all those RPs covered by this region are assigned to the first MS. When going from partition 2 to partition 3 in an anticlockwise direction, the RPs encountered are assigned to the second MS and so on. This process is illustrated in Fig. 5.6.

When going from a particular partition χ_i to partition χ_{i+1}, the following cases are possible:

Case 1: The number of RPs is greater than a threshold value (r_{th}), then the RPs are assigned to an MS.

Case 2: The number of RPs is less than r_{th} or equal to 0, then either χ_i is merged with χ_{i-1} or χ_{i+1} is merged with χ_{i+2} so that the number of RPs becomes acceptable in the region. In other words, this partition is annihilated with the next or the previous partition to make number of RPs greater than or equal to r_{th}.

Besides these cases, if by a particle all partitions $\chi_i^{\prime}s$ coincidentally are annihilated forming a single region, then the particle is modified appropriately. This is done as multiple subtours for MSs is desired rather than a single tour.

Each particle is evaluated using an objective function to decide the quality of the solution. The optimal number of MSs should be obtained such that the subtour lengths are balanced. Consequently, the delay in data collection is also minimized. Therefore, the objective function is a minimization function written as follows:

$$Minimize\ w_1\Gamma + w_2\Lambda \tag{5.37}$$

The smaller the value of the objective function, the better is the fitness of the particle. The pseudo-code of the proposed MSs determination method is given in Algorithm 9. Fig. 5.7 shows how the Jaya algorithm evaluates a group of particles to reach an optimal solution. The dimension of each particle is 4, which indicates that the maximum number of subtours possible is 4. On evaluation and refinement of the population, the optimal number of subtours turns out to be less than this maximum value. The number of subtours is finally designated as the optimal number of MSs.

Algorithm 9: Algorithm to determine optimal number of MSs

Input : R, D_{tour}
Output : v

Initialize the particle position ($X_{j,k,i}$) with a random number between 0 and 1
Initialize the parameters of Jaya algorithm
$l^{opt} \leftarrow$ Find the fitness of each particle ($X_{j,k,i}$)
$g^{opt} \leftarrow$ Find the particle with $max(l^{opt})$
for $iter = 1$ *to* max_{iter} **do**
  Update particles position ($X'_{j,k,i}$) using Eq. 5.36
  $l^{opt} \leftarrow$ Find the fitness value of particles ($X'_{j,k,i}$)
  if $fitness(g^{opt}) > max(l^{opt})$ **then**
    update g^{opt} with the fittest particle
  end
end
$D_{stour} \leftarrow$ Obtain the solution using the g^{opt} particle and D_{tour}
$D'_{stour} \leftarrow$ Apply the approximate shortest tour finding algorithm to each subtour in D_{stour}
$v \leftarrow |D'_{stour}|$

5.6 Simulation Analysis

5.6.1 Simulation Settings

We performed extensive simulation of our proposed algorithms using Python 3.5 on Ubuntu operating system (64 bit) having Intel(R) Core(TM) i7-8550U CPU 1.80GHz 8 processor and 16 GB RAM. We considered a network where SNs were placed in a $1000 \times 1000\ m^2$ monitored square area randomly and varied from 200 to 600. For the sake of simulation, two network configurations were created. In the first configuration (WSN#1), the SNs were distributed throughout the network by keeping the number of network segments fixed (say 10), whereas, in the second configuration (WSN#2), a fixed number of SNs were distributed in a varying number of network segments. The number of network segments was varied from 10 to 26. The number of SNs in the second configuration was fixed to 600. Each MS moves with a speed of 5 m/s for evaluation of different metrics considered. However, the speed of MSs was also varied from 5 $m/s - 9\ m/s$ for evaluation of data collection latency metric. The different parameters and their values used in the simulation are shown in Table 5.1.

The proposed data collection scheme is to assemble Algorithm 8 and Algorithm 9 in sequence. It is worth mentioning that in Algorithm 9, for dimension of a particle, we tried the simulation for different values of $\beta = 0.2, 0.3, 0.4, 0.5$. However, it was found that considering $\beta = 0.30$ was most effective.

We compare our scheme with the following popular tour planning algorithms, i.e.,

Table 5.1: Simulation parameters

Parameter	Value
Number of SNs (n)	200-600
Network size	$1000 \times 1000\ m^2$
Communication range (c^s)	20 m, 40 m
Network segments	10 - 26
Initial energy of SNs	30 J
Electronics energy (E_{elec})	50 nJ/bit
Packet size	4000 $bits$
Amplifier energy in free space model (ε_{fs})	10 $pJ/bit/m^2$
Amplifier energy in multi-path model (ε_{mp})	0.0013 $pJ/bit/m^4$
Data aggregation energy (E_{DA})	5 nJ/bit
Number of iterations (max_{iter})	300
Population size	100
BS location	$(500, 500)$
Average speed of a MS (v_m)	$5m/s - 9m/s$

MCPP [30], and STC [36]. The reason is that both these algorithms are similar to our proposed scheme in the sense that they also determine a set of RPs first and then obtain subtours for multiple MSs. Additionally, single-hop communication is considered in both these schemes. The parameters used for performance evaluation are as follows:

- *Number of RPs* is counted in a tour of a MS. While average RPs are the number of RPs per network segment.

- *Path length* is the summation of the euclidean distance travelled by all MSs to complete their associated tours. This is referred as $|D_{stour}|$ as per Eq. 5.15.

- *Number of MSs* is required to perform optimized data collection in terms of delay, path length, and network lifetime. The average MSs is the number of MSs required per network segment.

- *Total energy consumption* is the total amount of energy spent by the MSs to complete a tour, collect data at RPs, and communicate the same to the BS. It also includes the energy spent by SNs to send data to the MS when it reaches an RP.

- *Network lifetime* is the number of rounds until 70% of the SNs are functional or when all SNs at an RP is dead so that the path of a MS breaks, whichever happens earlier. A *round* is the time taken by a MS to complete its tour for data collection and communicate the data to the BS once. The average network lifetime is the network lifetime per network segment.

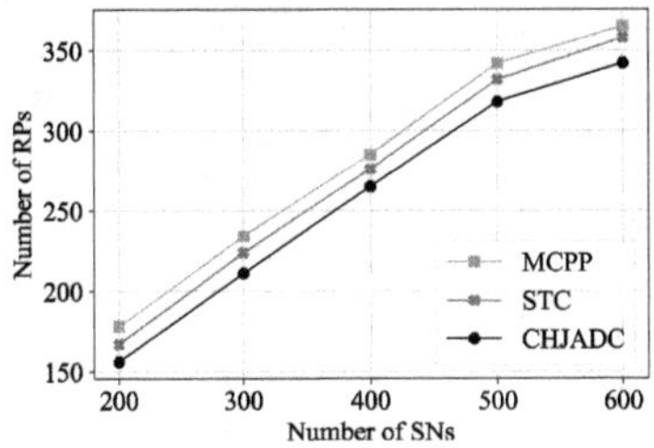

(a)

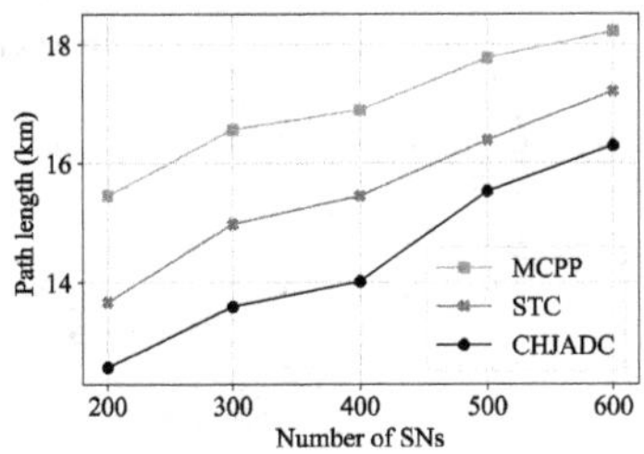

(b)

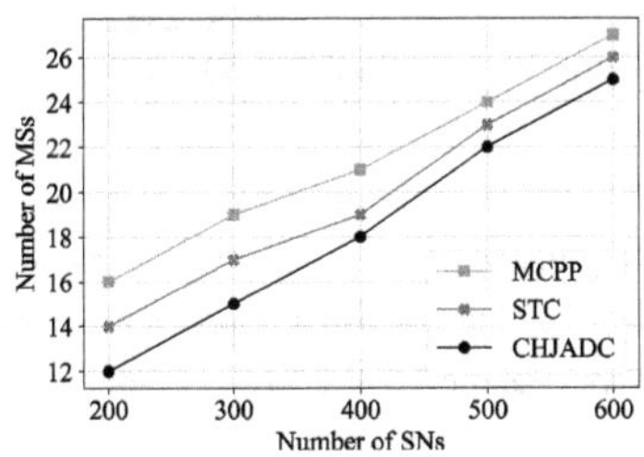

(c)

Figure 5.8: Performance comparison for WSN#1 of (a) Number of RPs vs number of SNs, (b) Path length vs number of SNs, (c) Optimal number of MSs corresponding to number of SNs.

- *Data collection latency* is the amount of time taken to gather data from all SNs by the MSs assigned to them in one round. Assuming the average speed of the MS to be v_m and path length to be L, the data delivery delay (T_{delay}) is computed by $T_{delay} = \frac{L}{v_m}$.

The simulations were carried out for each of the aforesaid parameters and the results were generated by taking an average of 20 runs. Similarly, the results were generated

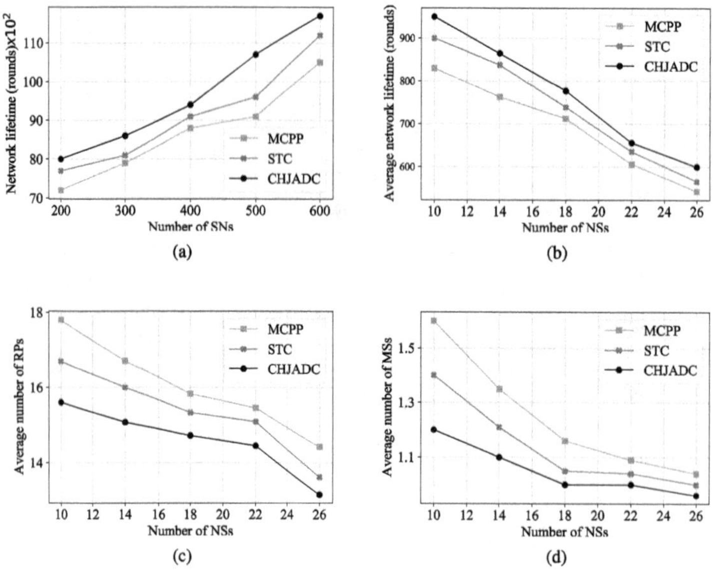

Figure 5.9: Performance comparison of (a) Number of SNs vs network lifetime (WSN#1), (b) Number of SNs vs average network lifetime (WSN#2), (c) Number of SNs vs average number of RPs (WSN#2), (d) Number of SNs vs average number of MSs (WSN#2).

by taking the average of 20 runs of STC and MCPP for the sake of comparison. For the evaluation of the number of RPs, path length, and network lifetime, a single MS has been used. Similarly, the optimal number of MSs obtained using Algorithm 9 has been used to evaluate average RPs, average network lifetime, total energy consumption, and data collection latency.

5.6.2 Number of RPs

The number of RPs is a vital parameter as it directly influences the computational effort of an algorithm [32]. Therefore, less number of RPs indicates less computational effort and vice versa. Fig. 5.8(a) demonstrates the influence of the varying number of SNs on the number of RPs. It can be observed that with the increase in the number of SNs, the number of RPs also increases. This is intuitive because more RPs are required to cover more SNs. However, in comparison to MCPP and STC algorithms, the proposed algorithm achieves a significantly lower number of RPs due to efficiently combining SNs to form RPs using the convex hull-based method, and thus it incurs significantly less computational effort. In other words, the proposed method can manage the topological organization of the SNs in an efficient way to find their representative locations (i.e., RPs) resulting in significantly lower number of RPs.

Fig. 5.9(c) shows the performance comparison in terms of average RPs. The proposed scheme achieves less average RPs than other schemes. Moreover, the figure also shows that average RPs increase with the increase in the number of SNs. The reason is as follows. With the increase in network segments, the same number of SNs gets sparsely distributed. Therefore to cover the same amount of SNs, more RPs are needed. Consequently, the number of RPs per network segment increases.

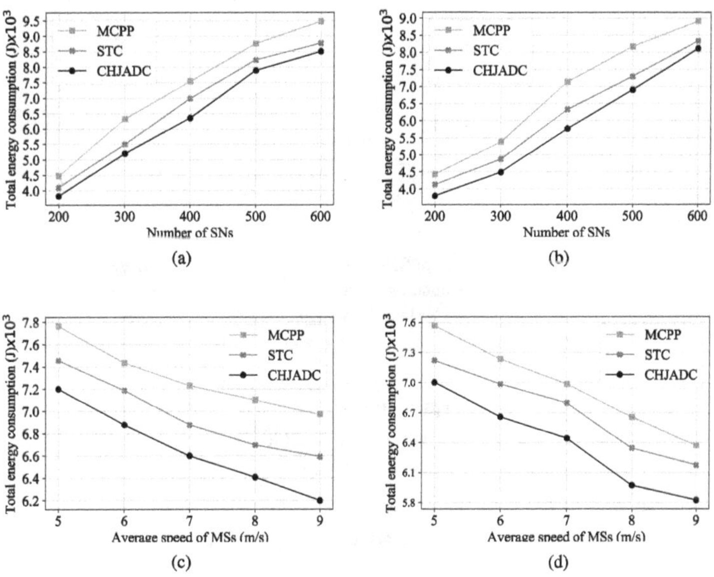

Figure 5.10: Performance comparison of (a) Total energy consumption vs number of SNs ($c^s = 20\ m$, WSN#1), (b) Total energy consumption vs number of SNs ($c^s = 40\ m$, WSN#1), (c) Total energy consumption vs average speed of MSs ($c^s = 20\ m$), (d) Total energy consumption vs average speed of MSs ($c^s = 40\ m$).

5.6.3 Path Length

Fig. 5.8(b) demonstrates the performance of the proposed scheme compared to other schemes in terms of the path length of the MSs. Note that path length directly influences

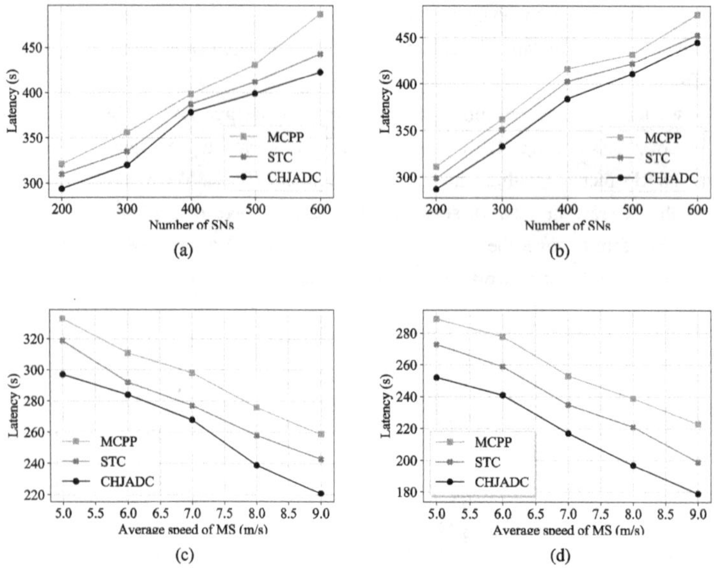

Figure 5.11: Performance comparison of (a) Number of SNs vs the corresponding latency for c^s = 20 m (WSN#1), (b) Number of SNs vs the corresponding latency for c^s = 40 m (WSN#1), (c) Average speed of MSs vs latency for c^s = 20 m (WSN#1), (d) Average speed of MSs vs latency for c^s = 40 m (WSN#1).

delay and energy consumption of data collection so that a shorter path length would enhance the efficiency of the same. From the figure, it can be observed that the proposed scheme generates a shorter path for each instance of the number of SNs. In Fig. 5.8(b), the path length increases with the increase in the number of SNs. This is reasonable as RPs increase with an increase in the number of SNs, and thus path length increases eventually.

5.6.4 Number of MSs

In Fig. 5.8(c), the performance of the proposed scheme is shown with a comparison to other schemes in terms of the number of MSs. It is noteworthy that a large number of MSs will incur high infrastructure cost, whereas a less number of MSs will lead to buffer overflow of SNs, longer path length, and data collection delay. In such situations, the scheme must have an optimal number of MSs to keep a balance between the above issues. From the figure, we can see that number of MSs in the proposed scheme is optimal than the other schemes.

In Fig. 5.9(d), the performance comparison is shown in terms of average MSs. The proposed scheme requires less average MSs than the other schemes. From Fig. 5.9(d), it is also observed that average MSs increases with the increase in the number of SNs. It is understood with the fact that with the increase in the number of SNs the number of RPs also increases; as a result, the number of MSs also increases to cover the RPs for optimized data collection.

5.6.5 Network Lifetime

The network lifetime and energy consumption per round are correlated with each other and directly influences the performance of the WSN. When the energy of the SNs is high, it leads to the increased network lifetime. Fig. 5.9(a) shows the network lifetime of the proposed scheme in comparison with other schemes. Here, as the SNs can reach MS in a single hop, no packet relay by SNs is needed. This explains why the network lifetime actually increases with the increase in number of SNs [36].

Fig. 5.9(b) represents the performance comparison in terms of average network lifetime. Our proposed scheme achieves a higher average network lifetime than the other schemes. It is also evident from the figure that the network lifetime decreases with the increase in the number of SNs. The reason is that with the increase in the number of SNs, the node density per segment decreases; as a result, the number of SNs per RP also decreases. This leads to the path break of the MS comparatively in less time, thereby making SNs die earlier.

5.6.6 Total Energy Consumption

Figs. 5.10(a) and (b) represent the performance of the proposed scheme in comparison with other schemes in terms of total energy consumption varying with the number of sensor nodes. From the figure, we can observe that the performance of the proposed scheme is high as compared to other schemes. In particular, the proposed scheme was able to achieve 13.67 % and 14.46 % improvement over MCPP scheme, and 5.7 % and 6.62 % improvement over STC scheme for 20 m and 40 m of communication range, respectively. The reason is that an efficient organization of the network in our proposed scheme leads to better energy utilization. Furthermore, it is observed that, with the increase in the number of SNs, the energy consumption per round increases. This is because more SNs will spend more energy (overall) for sensing data from the environment.

Figs. 5.10(c) and (d) show the performance of the proposed scheme in terms of total energy consumption varying with the average speed of the MSs. Here, we assumed that the number of SNs were fixed to 600 and the number of network segments was fixed to

10. From the figure, we can observe that all algorithms yield lower energy consumption when the MSs move faster. The reason for this is that when a MS travels faster (at a steady speed) it consumes lesser energy as it covers the whole tour in a lesser time. As a result the overall energy consumption is reduced [39, 114]. The proposed scheme yields the best performance among all the algorithms. In particular, the proposed scheme was able to achieve 8.86 % and 8.4 % improvement over MCPP scheme, and 4.41 % and 4.9 % improvement over STC scheme for 20 m and 40 m of communication range, respectively. This is because the convex hull formation phase of the proposed scheme organizes the topology of the network efficiently, thus leading to better energy utilization. Additionally, the optimal number of MSs plays a key role in the conservation of energy of the SNs, which might be frequently wasted in their sleep and wake-up cycles, as compared to other schemes.

5.6.7 Data Collection Latency

Figs. 5.11 (a) and (b) show the performance of the proposed scheme in terms of the total data collection latency varying with the number of SNs. Note that, the latency depends on number of RPs, speed of the MS, and the size of the sensed data. As these factors are directly affected by the number of sensor nodes, therefore we have shown the variation of data collection latency with number of SNs. From the figure, we can observe that the proposed scheme has the lowest latency. In particular, the proposed scheme achieves 6.92 % and 8.82 % improvement over MCPP, and 3.92 % and 3.64 % improvement over STC for 20 m and 40 m of communication range, respectively.The reason is that for the proposed scheme the number of RPs is less as compared to the other schemes, which results in a shorter path length. As the latency is directly proportional to the path length, hence the latency gets reduced.

Figs. 5.11 (c) and (d) show the total data collection latency with respect to average speed of the MSs. In this case, we assumed that the number of SNs were fixed to 600 and the number of network segments was fixed to 10. Here also, the proposed scheme wins as it is able to achieve 11.52 % and 15.53 % improvement over MCPP, and 5.86 % and 8.64 % improvement over STC for 20 m and 40 m of communication range, respectively. Each of the MSs moves with a constant speed on the path and collects data in one-hop from SNs along the path. As the data collection latency is directly proportional to the speed of the MSs, therefore with the increase in their speed, the data collection latency reduces. Also, using an optimal number of MSs creates a balance between the time and cost of the proposed scheme. The plots in Figs. 5.11 (c) and (d) testifies these expected outcome for this parameter.

Table 5.2: Results of ANOVA test

Performance metric	F-statistic	F-critical	p-value
Total energy consumption	92.58686541	3.219942293	4.0245E-16
Data collection Latency	21.73986972	3.219942293	3.3045E-07

5.6.8 Statistical Validation

In this section, we perform one way ANOVA [115] test for statistical validation of the simulation results. The following two cases are considered:

1. Null hypothesis (H_0): The means of different groups are assumed to be the same. We have considered the null hypothesis as "$H_0 : \mu_{MCPP} = \mu_{STC} = \mu_{CHJADC}$".

2. Alternate hypothesis (H_1):The means of different groups are assumed to be different. We have considered the alternate hypothesis as "$H_1 : \mu_{MCPP} \neq \mu_{STC} \neq \mu_{CHJADC}$".

The differentiation can be concluded by rejecting H_0. The conditions to reject H_0 are: 1) p-value is not higher than the significance level ($\alpha = 0.05$) and 2) F-statistic is greater than F-critical, where p-value, F-statistic, and F-critical are ANOVA test specific parameters. To conduct the test, we have considered the number of SNs to be 300, communication range to be 40 m, and the sample size to be 20. Further, we have considered two parameters total energy consumption and data collection latency for the same. The reason is that these parameters are paramount for sustainable nature of the proposed scheme and are proportional to the other parameters.

The results of ANOVA test are depicted in Table 5.2. From the table it can be observed that the aforementioned two conditions to reject H_0 are satisfied, i.e., F-statistic is greater than F-critical and p-value is not higher than α are satisfied. Thus, it can be concluded that H_0 is rejected. It means that the alternate hypothesis (H_1) is accepted. Thus, the means of the total energy consumption and data collection latency for MCPP, STC, and CHJADC are significantly different. We have also presented the 95% confidence interval of the means of the these schemes in terms of the the aforementioned parameters. In Fig. 5.12 we have shown the interval plots using the normal distribution. Here, μ_1, μ_2, and μ_3 represent the means of the aforesaid parameters for MCPP, STC, and CHJADC, respectively.

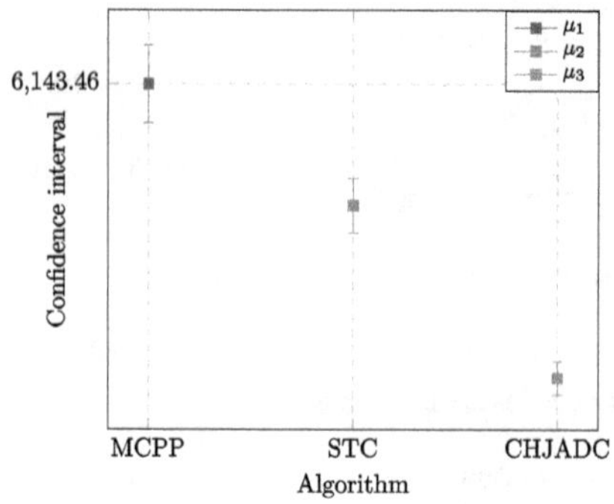

(a)

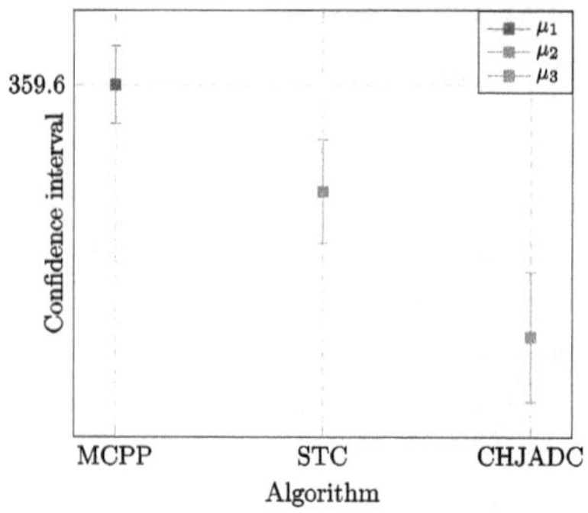

(b)

Figure 5.12: Comparison of confidence interval for MCPP, STC, and CHJADC in terms of (a) Total energy consumption and (b) Data collection latency.

5.7 Chapter Summary

In this chapter, we have proposed a novel data collection scheme to solve the data collection problem for disjoint WSNs with mobile edge nodes by considering a realistic propagation model of the SNs. The goal is to determine a set of RPs, an optimal tour of a single MS, and an optimal number of MSs with their balanced trajectories. The scheme is ideal for delay-tolerant applications requiring a shorter traveling path, but it is also suitable for connected networks. The scheme comprises two algorithms; the first one is based on a convex hull for determining a set of RPs and constructing an optimal tour of a single MS. The second one is based on the Jaya algorithm for finding an optimal number of MSs determined through a subtour optimization problem. Note that the better organization of the network via convex hull formation and an appropriate number of MSs contributes to the better performance of the proposed scheme. Extensive simulations have shown that the proposed scheme can boost data collection performance compared to other competing approaches in terms of various parameters. Statistical validation of the simulation results through the ANOVA test has also been carried out along with plotting of confidence interval. However, the scheme may not be suitable for obstacle-ridden WSNs. Thus, in the next chapter, the presence of obstacles is exploited in data collection using a single MS.

Chapter 6

Obstacle Adaptive Data Collection in the Internet of Things

Chapter 6

Obstacle Adaptive Data Collection in the Internet of Things

6.1 Introduction

Similar to the previous chapter, this chapter considers the edge-based Internet of Things (IoT), wherein wireless sensor networks (WSNs) are the predominant channel for data dissemination that employ mobile edge nodes such as mobile sinks (MSs). However, MSs may face obstacles, which become hurdles in data collection. In this regard, the existing works [30, 77, 78, 83] have the following issues. 1) How to detect the position of RPs in convex part of the obstacles and how many RPs (overall) will be there in the network, 2) How to re-route the MS for its seamless movement around the obstacle, and 3) How to remove sharp turns in the path of the MS. In other words, they have neglected smooth path construction, i.e., the path must be obstacle-avoiding that combines the properties of being both smooth and fast to traverse [84] so that delay and energy consumption will be significantly reduced. The proposed scheme address all the aforesaid issues.

The path planning problem is first formulated to observe the properties of the path to be smooth and fast. Then, the problem is proved to be NP-complete. Next, we present the proposed scheme, which is divided into two phases as follows. The first phase determines a set of RPs using a polynomial-time solution to the set cover problem and finds a tour for the MS using an approximation algorithm. In the second phase, a combination of the Bezier curve and modified BUG2[66] algorithm is used for obstacle avoidance and smooth path construction. The proposed scheme is called **O**bstacle **A**ware **S**mooth **P**ath **P**lanning (OAsPP) that efficiently designs the trajectory of the MS for data collection in the presence of obstacles.To the best of our knowledge, this is not considered by the existing works. We perform extensive simulations of our proposed work and compare the results with the existing algorithms to exhibit its superior performance in terms of various performance metrics. We also performed statistical validation of the simulation results.

6.2 Preliminaries

6.2.1 Important Definitions

Definition 6.2.1.1 (*Obstacle*): An obstacle is an entity that may have irregular shape and must be avoided by an MS. Their shapes can be estimated by a convex/concave polygon or both.

Definition 6.2.1.2 (*Smoothness of a path*): A path is said to be smooth if it is tangent continuous, has a continuous curvature, or has a continuous derivative of curvature [84].

Definition 6.2.1.3 (*Path*): A path of an MS is the route it traverses for data collection through the deployment region by moving from one RP to the other RP. The starting and ending location of the MS is BS (i.e., r_0). It is expressed mathematically as follows:

$$D_{tour} = < r_0, r_1, r_2, \ldots, r_m, r_0 > \tag{6.1}$$

and the length of the tour is

$$|D_{tour}| = L = \sum_{i=1}^{n^r-1} d_{i,i+1} + d_{0,1} + d_{n^r,0} \tag{6.2}$$

where $d_{i,j}$ is the euclidean distance between r_i and r_j.

Definition 6.2.1.4 (*Smoothness measure*): Smoothness measure is the integral over the square of arc-length derivative of curvature along the path. The rationale of using this measure is that it is proportional to the jerk which is unwanted for a vehicle. It also taints the speed of a vehicle. Let $f(x)$ be a function having curvature $K(x)$, $x \in [x_o, x_1]$, which can be parameterized in its arc length s, then the cost function is expressed as follows [116]:

$$\int_{s(x_o)}^{s(x_1)} \dot{K}(s^2)ds = \int_{x_0}^{x_1} \frac{K'(x^2)}{\sqrt{1+f'(x^2)}}dx \tag{6.3}$$

where $\dot{K} = \frac{dK}{ds}$ and $K' = \frac{dK}{dx}$

Definition 6.2.1.5 (*Bezier curve*): A Bezier curve is a parametric curve expressed mathematically as follows:

$$P(t) = \sum_{i=0}^{n} b_i B_i^n(t) \tag{6.4}$$

where b_i's are the control points and $B_i^n(t)$ is called the Bernstein polynomial or the basis function. Here, $B_i^n(t) = \binom{n}{i} t^i (1-t)^{n-i}$ and $\binom{n}{i} = \frac{n!}{i!(n-i)!}$. Note that the Bezier curves are potentially capable of creating smooth paths for vehicles [117].

Definition 6.2.1.6 (*Intersection control points*): Intersection control points (ICPs) are the points obtained by the intersection of the path of an MS with the obstacles.

Definition 6.2.1.7 (*Potential rendezvous points (PRPs)*): It is the set of all RPs used in the process of determination of final set of RPs.

6.2.2 System Model

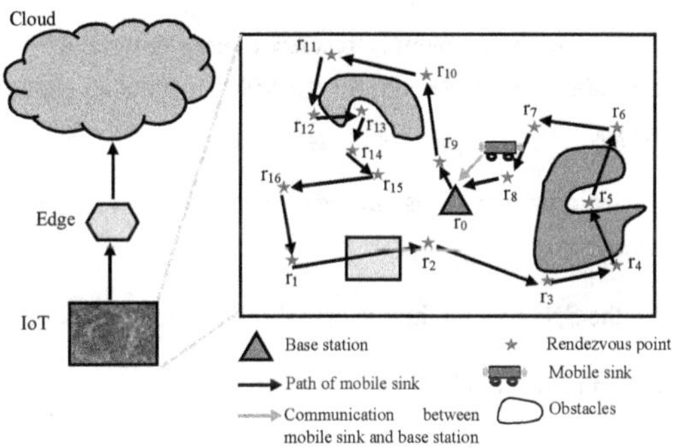

Figure 6.1: An example of three layer architecture involving edge. The IoTs are essentially WSNs deployed at the edge of the network.

The system model here is similar to the previous chapter, with minor differences. However, for the sake of readability, we again detail the model as follows. Fig. 6.1 shows three layers of a sample architecture where cloud servers are at the top layer, edge servers are at the middle layer, and WSNs are at the bottom layer. Different WSNs have different mobile edge nodes (MS) to collect data and send it to the cloud servers for further processing. The edge servers are capable of pre-processing the received data from the MS. In each WSN, there is a set of SNs, $S = \{s_1, s_2, \ldots, s_{n^s}\}$, a set of RPs, $R = \{r_1, r_2, \ldots, r_{n^r}\}$, a BS r_0, and an MS. The BS is aware of the location of the SNs and MS. Also, the MS is aware of the location of the SNs and the BS. Note that the set R is independent of S. In other words, an SN may or may not be present at an RP. The SNs are homogeneous, so they have the same buffer size and communication range. Each SN s_i has a sensing rate r_i^s

(i.e., the number of packets produced per unit of time), and it can store at most n^p sensed packets in its buffer. The MS periodically visits the RPs in the network to collect data from SNs. It offloads this data to r_0, which sends it to edge servers that in turn offloads this data to cloud servers. It has sufficient storage capacity and energy to complete the tour in time. The speed (v) of the MS is sufficient to prevent the buffer overflow problem. The time taken at an RP for gathering data is negligible. The MS also performs aggregation of data at each of the RPs. The energy model used in communication between SNs and MSs is the same as in [18]. Different from the traditional data collection models, 1) individual calculation for each node is no longer needed. The calculation needs to be done at the mobile node (i.e., MS). 2) The path around the obstacles is smooth, resulting in a faster traversal of the MS.

6.3 Problem Formulation

Given the set of SNs (S), the path planning problem is to select a set of optimal RPs R in the presence of obstacles and decides a traveling path for the MS that starts and ends its journey at r_0 for collecting data such that:

1. the length of the obstacle-avoiding path (L) is minimized so that this path is smooth and fast (refer to Definition 2),
2. the energy expenditure of the SNs to send their data to the MS is minimized,
3. the latency of data collection is also minimized.

We refer to the aforesaid problem as modified optimal path planning (MOPP) problem.

In the constraint 1, the Bezier curve ($P(x)$) associated with suitable points on the path D_{tour} and ICPs must satisfy the following conditions to ensure the smoothness of the path.

1. The endpoint constraints are satisfied by $P(x)$, and the endpoints are chosen suitably from D_{tour} and ICPs.
2. $P(x)$ maximizes the cost function

$$\int_I \frac{K'(x^2)}{\sqrt{1+P'(x^2)}}dx \tag{6.5}$$

   where $K(x) = \frac{P''(x)}{(1+P'(x)^2)^{\frac{3}{2}}}$, $I = [x_0, x_1]$, and x_0 and x_1 are the two extreme points of the curvature.

Theorem 1: The MOPP problem is NP-complete.

Proof. We first show that the MOPP problem belongs to the NP class as follows. Given an instance of the MOPP problem and path length (L), we can verify whether the path is feasible in polynomial time. Hence, it belongs to the NP class.

Next, we reduce the Hamiltonian cycle problem to the MOPP problem. Note that the former is an NP-complete problem [118]. Given a set of n points, the minimum Hamiltonian cycle decision problem poses the question whether a cycle of length L exists that passes through each of these points exactly once except the starting point. Now, we state the MOPP decision problem as follows:

Given S and r_0, the MOPP decision problem asks whether there exists a traveling path having length L for the MS to gather data from all SNs with total energy consumption E^t and data collection latency (delay) D.

Let us assume that each SN on average produces n^p packets. Therefore, the total energy consumption will be

$$E^t = n^p n^s (E'_{RX} + E'_{TX}) \tag{6.6}$$

where E'_{RX} and E'_{TX} are the amount of energy consumed to transmit and receive a packet, respectively. Thus E^t is also the total energy used by SNs for communication. Note that, for simplicity we have neglected components of energy consumption such as energy spent for data aggregation, energy spent by MS for movement, etc. Now, let us assume that the MS moves with a speed v m/s, then the total data collection latency is

$$D = \frac{L}{v} \tag{6.7}$$

to visit $n^r + 1$ points (i.e., n^r RPs and 1 r_0). Note that for simplicity we have again neglected other components of latency such as data collection time at an RP, packet size, etc. Here, the MOPP problem instance asks to find a Hamiltonian cycle of length $L = Dv$ by covering points $n^r + 1$ points with the total energy expenditure E^t which is similar to the Hamiltonian cycle decision problem. This reduction takes polynomial time, thus MOPP is also NP-complete. □

6.4 The Proposed Scheme: OAsPP

The scheme has two phases: 1) In the first phase, the SNs present in the network are organized into a set of an optimal number of RPs, and a path is determined using a heuristic algorithm that goes through obstacles and is not smooth, and 2) In the second phase, the current path is modified to make it smooth using another heuristic algorithm that combines Bezier curve (obtained using quadratic interpolation) and modified BUG2 algorithm. The objective of the work is to minimize the path length while adding the

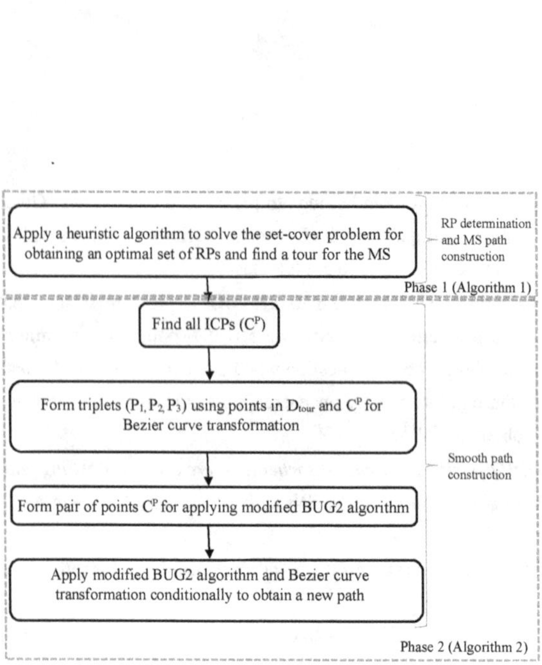

Figure 6.2: An outline of the proposed scheme.

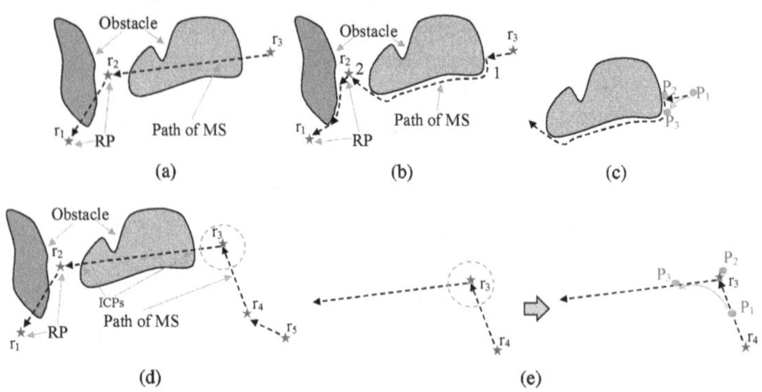

Figure 6.3: Example of constructing a smooth path (a) Path of an MS through two different obstacles, (b) Path of an MS avoiding obstacles (1 and 2 represent sharp turns), (c) Smooth path construction at an ICP by assuming three control points, i.e., P_1, P_2, and P_3, (d) Segment of path with sharp turn (red broken circle) and ICPs, (e) Smooth path construction at a site which is not ICP but has a sharp turn using three control points, i.e., P_1, P_2, and P_3.

property of being smooth and minimize energy consumption and data collection latency to enhance the network lifetime. The outline of the proposed scheme is represented via a flowchart in Fig. 6.2. The phases are detailed in the following sections.

Algorithm 10: Algorithm to determine optimal set of RPs and the tour of MS

```
Input  : ℵ
Output : R, D_tour
P_f ← Set cover, initially empty
P^p ← ℵ
p_i^size ← Compute the size of every element p ∈ P^p, and store them in size(p)
S' ← Set of SNs covered, initially empty
R_A ← Array of length max(size(p), p ∈ P^p) consisting of empty lists
for p ∈ P^p do
    Add p to R_A[size(p)]
end
I_max ← Index of the largest nonempty list in R_A
L ← Array of length n^s, initially empty
for p ∈ P^p do
    for k ∈ p do
        Add p to L[k]
    end
end
while I_max > 0 do
    p_0 ← any element of R_A[I_max]
    Add p_0 to P_f
    Remove p_0 from R_A[I_max]
    for k ∈ p_0 \ S' do
        for p ∈ L[k] do
            Remove p from R_A[size(p)]
            size(p) ← size(p) − 1
            if R_A[I_max] == φ then
                I_max ← I_max − 1
            end
        end
        Add k to S'
    end
end
R ← Find Centroid of all sets in P_f
P^M ← Find the approximate shortest tour using points in P_fin and R [21]
```

6.4.1 Determining RPs and Finding Tour of the MS

The problem of determining RPs is solved by mapping it to a set-cover problem. Note that the problem is NP-hard in nature [118]. At first, an adjacency matrix (a) is formed whose entries are as follows:

$$a_{i,j} = \begin{cases} 1 & \text{if } s_i \text{ is in communication range } c^s \text{ of } s_j, \\ 0 & \text{otherwise.} \end{cases} \tag{6.8}$$

This matrix is processed to form sets. Let us assume that there are 11 SNs (i.e., $s_1, s_2, \ldots, s_{11}$). Now, if s_1 is within communication range of SNs s_2, s_4, s_6, and s_7, then the first set will be $\{1,2,4,6,7\}$. If s_2 is within communication range of SNs s_3, s_7, s_9, s_{10} and s_{11} then, the second set will be $\{2,3,7,9,10,11\}$. However, if s_3 is within communication range of SNs s_2, s_7, s_9, s_{10}, and s_{11}, the third set will be $\{2,3,7,9,10,11\}$, which is identical to the second set. Thus, the third set is not considered for processing and is discarded. In this way, other distinct sets called as the potential sets or potential RPs (PRPs) are formed. These PRPs are input to Algorithm 10, which are processed using a greedy approach to obtain the set-cover (i.e., final set of RPs R). The formal description of the whole process is stated in Algorithm 10. The centroid of all the sets in set-cover are designated as RPs which is then used to form a tour of the MS. Note that the statements inside the for loops run in constant time. So, the time complexity of this algorithm (Steps 1-30) is $O(\sum_{p\in\aleph}|p|)$ and the time taken to compute the approximate shortest tour is $O((n^r)^2)$. Thus, the total time complexity is $O(\sum_{p\in\aleph}|p|+(n^r)^2)$.

6.4.2 Constructing a Smooth Path

The path obtained in the previous phase has sharp turns at some points and also at ICPs. For example sharp turns are shown in Fig. 6.3 (b) with the labels 1 and 2. These turns create problems for MS, hampering the internal hardware to some extent. The speed of the vehicle also reduces at these turns, which affects the energy consumption and data collection latency. To overcome these issues, the following parts need to be modified: 1) The portion of the path which is affected by obstacles and has a "sharp turn", 2) The portion of the path which is not affected by obstacles, but has sharp turns. In both these case, the question is how to detect a "sharp turn" and then replace that portion of the path with a Bezier curve. Now, assuming that that any portion of the curve (path) can be approximated using three control points (P_1, P_2, and P_3), the equation is

$$c(t) = x(t)\hat{i} + y(t)\hat{j}. \tag{6.9}$$

where $\hat{i}$ and $\hat{j}$ are unit vectors along X and Y directions. The arc length (s) of this curve ($c(t)$) between the two control points (P_1 and P_3) is

$$s = \int_{P_1}^{P_3} \sqrt{\left(\frac{dx}{dt}\right)^2 + \left(\frac{dy}{dt}\right)^2}\, dt \tag{6.10}$$

The condition for a "sharp turn" of a curve is as follows: The curvature of the circle through (P_1, P_2, and P_3) is four times the area of the triangle formed by these points divided by the product of the three sides. Here, the curvature is the extent of deviation from the straight line. So, mathematically it is expressed as follows.

$$k = \left| \frac{d^2c}{ds^2} \right| \tag{6.11}$$

where k is the magnitude of the second derivative of the curve at the given point. The portion of the path which satisfies the above condition is replaced with a Bezier curve obtained using the quadratic interpolation as

$$B = (1-t)^2 P_1 + 2(1-t)tP_2 v + t^2 P_3 \tag{6.12}$$

where $t \in [0, 1]$ is a constant. Note that this process is applied to all parts of the curve, which satisfies the above-mentioned condition. Now, we present the method to move around the obstacle. The algorithm used for this is a modified version of BUG2[66]. The MS initially moves along the straight line r_3; r_2, which is called the reference line, until it hits an obstacle at some hitting point 1 (Fig. ??(b)); then, the MS moves along the boundary of the obstacle using the righthand rule (i.e., the right hand maintains contact with the obstacle) until it again reaches back to the reference line at some point 2 (Fig. 6.3(b)). Now, if the MS finds that it is closer to r_2 than from 1 and that it can make progress on the reference line, then it continues its straight-line walk towards r_2; otherwise, it continues to navigate around the obstacle. The above procedure iterates until the MS reaches r_2. Note that at the hitting points (ICPs) 1 and 2, the BUG2 algorithm is modified for smooth path construction. The algorithm produces a path of length at most $P^M + \sum_{i=n^o} \frac{n_i l_i}{2}$, where P^M is the tour length, n_i is the number of times the reference line crosses the ith obstacle, and l_i is the perimeter of the ith obstacle. For convex obstacles, BUG2 is essentially optimal. Figs. 6.3(b) and 6.3(c) demonstrate an example of this algorithm. The complete process of smooth path construction is formally stated in Algorithm 11 and Fig. 6.3.

Theorem 2: The total energy consumption is reduced by a factor $n^f \times n^{st}$ by replacing n^{st} sharp turns in the path of the MS by a Bezier curve, where n^{st} is the number of sharp turns in the path.

Proof. The total energy consumption is directly proportional to the path length traveled by the MS. Therefore, first we show that path length is reduced by a factor n^f and then generalize it to n^{st} sharp turns.

Fig. 6.3 (e) shows the three control points (P_1, P_2, P_3) at an RP r_3 at which the path of

Algorithm 11: Algorithm for avoiding obstacles and constructing smooth path

Input : D_{tour} (Path obtained using Algorithm 8)
Output : D'_{tour} (Modified path)

```
Form triplets (P1, P2, P3) sequentially using the points in D_tour and C^P
n^t ← Number of triplets
ℙ ← Find all pairs of RPs (path segments) which intersects with obstacles using C^P
for each pair p in ℙ do
    Apply the modified BUG2 algorithm to p
end
for i = 1 to n^t do
    if arc of path at i^th point has sharp turn then
        Replace the associated segment with a Bezier Curve
    end
end
```

MS has a sharp turn. The initial distance traveled by the MS here is

$$\begin{aligned} l^{init} &= \overline{P_1P_2} + \overline{P_2P_3} \\ &= \sqrt{(x_2-x_1)^2+(y_2-y_1)^2} + \sqrt{(x_3-x_2)^2+(y_3-y_2)^2} \end{aligned} \tag{6.13}$$

where (x_i, y_i) are the coordinates of the control points P_1, P_2, P_3, $1 \leq i \leq 3$.

The distance traveled by the MS due to replacement of the path by the Bezier curve is also the arc length (l^a) of the Bezier curve from the points P_1 to P_3. Note that the sharp turn has been replaced by the Bezier curve, so it becomes smooth. Additionally, the curve of the path becomes differentiable and its derivative is continuous.

Now, the equation of the Bezier curve using the three control points (P_1, P_2, P_3) is given by

$$P(t) = P_1t^2 + 2P_2t(t-1) + P_3(t-1)^2 \tag{6.14}$$

and the arc length of $P(t)$ from P_1 to P_3 is given by

$$l^a = \int_{P_1}^{P_3} \sqrt{1+[P'(t)]^2 dt} \tag{6.15}$$

where $P'(t)$ is the derivative of $P(t)$.

On solving the above equation and substituting the value of coordinates, we get:

$$l^a = \frac{(c_t^1 \times c_t^2) - c_t^3 \times \log(c_t^1 + c_t^2) - (v \times c_t^4 - c_t^3 \log(v + c_t^4))}{8 \times u^{\frac{3}{2}}} \tag{6.16}$$

where $c_t^1 = 2\sqrt{u(u+v+w)}$, $c_t^2 = 2u+v$, $c_t^3 = vv - 4uw$, $c_t^4 = 2\sqrt{uw}$, $u = 4((k_3)^2 + (k_4)^2)$, $v = 4(k_1k_3 + k_2k_4)$, $w = (k_1)^2 + (k_2)^2$, $k_1 = 2(x_2 - x_1)$, $k_2 = 2(y_2 - y_1)$, $k_3 =$

$x_3 - 2x_2 + x_1$, and $k_4 = y_3 - 2y_2 + y_1$.

Finally, the path length at this sharp turn is reduced by the factor, n^f, as follows.

$$n^f = \frac{l^a}{l^{init}} \tag{6.17}$$

Generally speaking, as there are n^{st} sharp turns and at each turn the path length is reduced by the factor n^f. It can also be deduced from above equation that $l^a < l^{init}$. Thus, the total reduction in path length is by the factor $n^f \times n^{st}$. As mentioned earlier, total energy consumption is directly proportional to the path length, i.e., $E^{tot} \propto |D_{tour}|$. Therefore, total energy consumption is also reduced by the factor $n^f \times n^{st}$. □

6.5 Performance Evaluation

6.5.1 Simulation Setup and Base Algorithms

This section evaluates the performance of the proposed scheme using simulation analysis. All the simulations were performed using Python 3.5 on Ubuntu operating system (64 bit) having Intel(R) Core(TM) i7-8550U CPU 1.80GHz 8 processor and 16 GB RAM. The network parameters assumed for the simulations are given in Table 6.1.

Table 6.1: Simulation parameters

Parameter	**Value**
Number of SNs (n)	100-300
Network size	$500 \times 500\ m^2$
Communication range (c^s)	20 m, 30 m
Initial energy of SNs	30 J
Electronics energy (E_{elec})	50 nJ/bit
Packet size (l)	4000 $bits$
Amplifier energy in free space model (ε_{fs})	10 $pJ/bit/m^2$
Amplifier energy in multi-path model (ε_{mp})	0.0013 $pJ/bit/m^4$
Data aggregation energy (E_{DA})	5 nJ/bit
BS location (r_0)	$(250, 250)$
Average speed of an MS (v)	9 m/s

We compare our scheme with the following competing path planning schemes, i.e., HOAA [83 [illegible] OACE [78]. The reasons for choosing these two base algorithms are as [illegible] ar to our proposed scheme, they also determine a path for the MS in the [illegible] es. 2) They also have dealt with obstacles which are static and irregular [illegible] e of comparison, the simulation results of 20 runs were averaged and [illegible] ferent performance metrics were plotted. We now present the results [illegible] tive performance metrics, i.e., the number of RPs, path length, total energy consumption, network lifetime, the standard deviation of RPs, relative smoothness of the path, and relative latency of data collection.

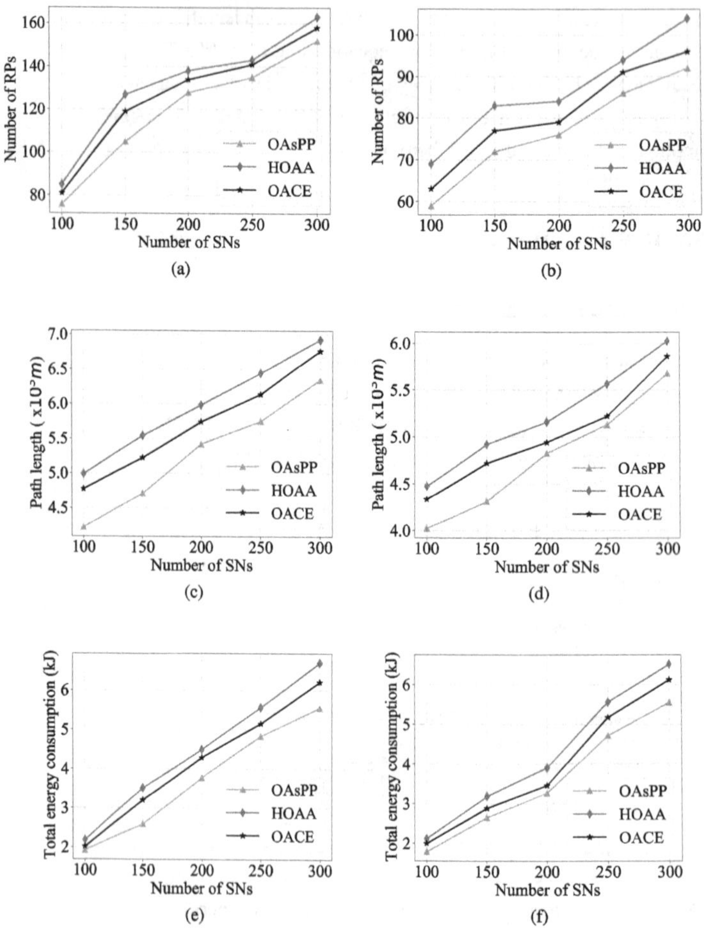

Figure 6.4: Performance comparison of number of SNs vs (a) Number of RPs ($c^s = 20\ m$), (b) Number of RPs ($c^s = 30\ m$), (c) Path length ($c^s = 20\ m$), (d) Path length ($c^s = 30\ m$), (e) Total energy consumption ($c^s = 20\ m$), (f) Total energy consumption ($c^s = 30\ m$).

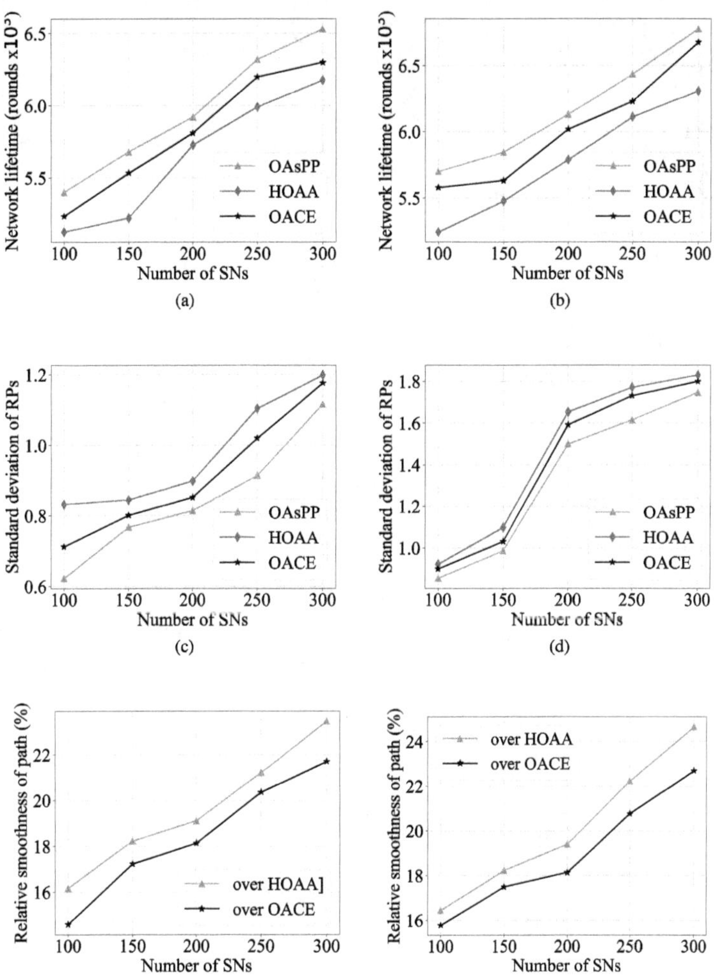

Figure 6.5: Performance comparison of number of SNs vs (a) Network lifetime ($c^s = 20\ m$), (b) Network lifetime ($c^s = 30\ m$), (c) Standard deviation of RPs ($c^s = 20\ m$), (d) Standard deviation of RPs ($c^s = 30\ m$), (e) Relative smoothness of path ($c^s = 20\ m$), (f) Relative smoothness of path ($c^s = 30\ m$).

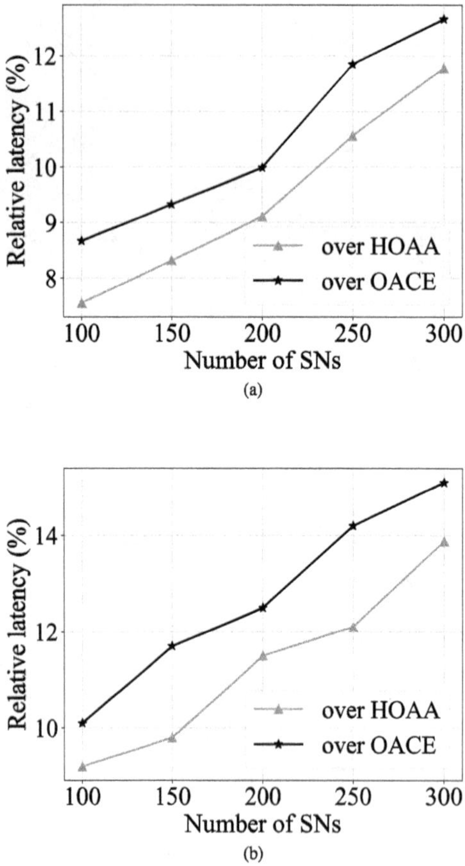

Figure 6.6: Performance comparison of number of SNs vs (a) Relative latency for data collection ($c^s = 20\ m$), (b) Relative latency for data collection ($c^s = 30\ m$).

6.5.2 Results

In Figs. 6.4(a) and 6.4(b), we compared OAsPP with HOAA and OACE algorithms in terms of number of RPs. It can be observed from the plots that with the increase in the number of SNs, the number of RPs increases. It is obvious to note that more sojourn locations are required for more SNs. However, OAsPP has a significantly lower number of RPs. In particular, OAsPP was able to achieve 10.48 % and 5.68 % improvement over

HOAA and OACE, respectively. Figs. 6.4(c) and 6.4(d) show the comparison of OAsPP in terms of the path length of the MS. We observe from the plots that the path length increases with the increase in the number of SNs. The reason is that with the increase in the number of SNs, the number of RPs increases, which is analogous to the traveling salesperson problem. This eventually increases the path length. Nevertheless, our scheme has the lowest path length. It achieves 10.12 % and 6.22 % improvement over HOAA and OACE, respectively. In Figs. 6.4(e) and 6.5(a) OAsPP is compared with other schemes in terms of total energy consumption. A similar trend can be observed, as mentioned previously, for the number of RPs and path length. The reason is that with the increase in path length, the MS has to spend more energy to move and collect data. Note that the energy consumption is the sum of energy consumption in packet processing, receiving data from SNs, aggregation of data, transmitting the data to the BS, and traveling along the path. In this case, also OAsPP wins with the lowest energy consumption by achieving 16.23 % and 9.39 % improvement over HOAA and OACE, respectively.

Next, the comparison of OAsPP with other schemes in terms of network lifetime is shown in Figs. 6.5(b) and 6.5(c). We assumed network lifetime as the number of rounds until 70% of the SNs are functional, where a *round* is the time taken by an MS to complete its tour for gathering data collection and communicating it to the BS once. The observation here is that with the increase in the number of SNs, the Network lifetime increases. This is evident as the node density increases with the increase in the number of SNs. Thus, that part of the network remains functional for a longer period of time. OAsPP achieves the best performance here with respect to the other schemes. It was able to achieve 6.29 % and 2.6 % improvement over HOAA and OACE, respectively. Figs. 6.5(d) and 6.5(e) shows the effect of number of SNs on standard deviation of RPs. Note that, lower the value of this metric, the better is the performance of the scheme. This parameter is indicative that the number of SNs associated with an RPs is balanced. With the increase in the number of SNs, the value of this parameter also increases. Our scheme achieves a relatively lower standard deviation value indicating a better distribution of SNs per RP. Here, OAsPP achieves 10.71 % and 6.14 % improvement over HOAA and OACE, respectively.

In Figs. 6.6(a) and 6.6(b), the comparison of OAsPP with other schemes in terms of relative smoothness of the path is shown. The metric used here to measure smoothness is given in Eq. 6.5. Note that HOAA and OACE algorithms don't have any mechanism to construct a smooth path. However, there are portions of the path that are naturally smooth and need no modification. Thus, we compared these schemes in terms of this parameter. It can be observed from the plots that the smoothness of the path increases with the increase in the number of SNs. This is because with the increase in the number of SNs, the number of RPs increases, and thus, the path length increases. In general, a path with more points on it is smooth as in traveling salesperson problem. Here, our scheme achieves better improvement over HOAA as compared to the OACE algorithm. Finally, in Figs. 6.6(c) and 6.6(d) the comparison of OAsPP with other schemes in terms relative latency of data collection is shown. The formula used for this parameter is given in Eq. 6.7. The data collection latency depends on the number of RPs, speed of the MS, the size of the data of the sensor nodes, and data processing capability of the MS. It can be observed that with the increase in the number of SNs, the latency also increases. This is justified as the major component of data collection delay is the traveling time of the MS, given the low speed of the MS. However, our scheme (OAsPP) achieved significantly lower latency due to smooth path construction. A smooth path naturally embeds all benefits associated with it into OAsPP. Again, our scheme achieves better improvement over HOAA than the OACE algorithm.

6.5.3 Analysis

We have observed that OAsPP outperforms HOAA and OACE algorithms in terms of the various metrics considered. The reason for this is that in the first phase, the topographical arrangement of SNs is efficiently managed to find sojourn locations. The first benefit was a significantly reduced number of RPs. The other two parameters, such as path length and energy consumption, are directly affected by the number of RPs and, in general, have a direct proportionality relation. It means that the increase in one will also increase the other. The second benefit was the enhancement of network lifetime and standard deviation of RPs. Generally speaking, network lifetime is inversely proportional to energy consumption. In OAsPP, the energy consumption was reduced, as mentioned earlier, thus the network lifetime was improved. The better performance in the standard deviation of RPs is also a consequence of the first phase of the scheme. The application of the Bezier curve and modified BUG2 algorithms to the second phase of the scheme explains the superior performance of OAsPP with respect to the relative smoothness of path and relative latency.

6.5.4 Results Validation

Here, we perform statistical validation of the simulation results using one way ANOVA test [115]. The test is associated with the following two cases: a) Null hypothesis (H_0), i.e., $H_0 : \mu_{OAsPP} = \mu_{HOAA} = \mu_{OACE}$ and b) Alternate hypothesis (H_1), i.e., $H_1 : \mu_{OAsPP} \neq \mu_{HOAA} \neq \mu_{OACE}$. Here, $\mu_{(.)}$ denotes the mean of a group. It concludes that the means of different groups are different by accepting H_1 and rejecting H_0. The conditions to

Table 6.2: Results of ANOVA test

Performance metric	F-statistic	F-critical	p-value
Total energy consumption	389.4763524	3.158842719	5.75778E-34
Path length	42.69676774	3.158842719	4.65544E-12

accept H_1 are: 1) p-value is less than the significance level ($\alpha = 0.05$) and 2) F-critical is smaller than F-statistic, where p-value, F-statistic, and F-critical are the parameters related to the test. In order to perform the test, the number of SNs is considered to be 200 and the communication range to be 30 m. We have considered path length and total energy consumption as these parameters most relevant to perform the test and also related to the objective of the scheme.

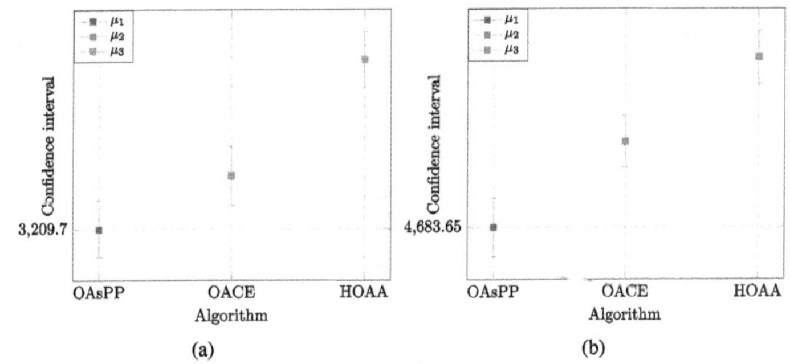

Figure 6.7: Comparison of confidence interval for OAsPP, OACE, and HOAA in terms of (a) Total energy consumption and (b) Path length.

Table 6.2 depicts the results of ANOVA test. It can be observed from the table that the conditions to accept H_1 are satisfied; thus, it can be concluded that H_1 is accepted. In other words, the means of the total energy consumption and path length for OAsPP, HOAA, and OACE are significantly different from each other. Fig. 6.7 shows the 95% confidence interval of the means of these schemes.

6.6 Chapter Summary

In this chapter, we have addressed the path planning problem and proposed a novel scheme by considering a greedy approach solution to the set cover problem, which runs in polynomial time, Bezier curve, and a modified BUG2 algorithm. The main consideration here is to plan a tour for an MS such that the path length is minimized and, at the same time, it is smooth. Additionally, the energy consumption and data collection latency are also minimized. We have also shown that OAsPP achieves performance gain over other relevant competing approaches in terms of various performance metrics through extensive experiments. In particular, OAsPP has been shown to achieve 10.48 % and 5.68 % improvement in terms of number of RPs and 10.12 % and 6.22 % in terms of path length over the competing approaches HOAA and OACE, respectively. It has also been shown that there has been an improvement of 16.23 % and 9.39 % in terms of total energy consumption and 6.29 % and 2.6 % with respect to network lifetime. The simulation results are validated statistically through ANOVA and posthoc analysis.

Chapter 7

Conclusions and Future Research

Chapter 7

Conclusions and Future Research

In this chapter we conclude the thesis by summarizing the entire work and presenting the concluding remarks. Thereafter, we present the future research directions for possible exploration.

7.1 Concluding Remarks

This thesis investigated the one-hop data collection problem using MS in different WSNs scenarios. At first, the background and motivation pertaining to various components of WSNs were presented in chapter 1. The problem considered for probing and then why it needs to be probed was explained. Next, a brief discussion about the contributions of our work as to what problems were solved how they were solved was presented. Next, in Chapter 2 the detailed study of the existing and competing works have been presented. The categorization of these works also has been done based on the general characteristics of the algorithms involved.

In Chapter 3 we have investigated the tour planning problem by employing an MS for data collection in the WSNs, which is suitable for applications where sensors are only partially connected. Hereby, we have first reduced the problem size by dividing the network into MUs by using BFS traversal and Tarjan's algorithm. Then, in order to obtain an optimal number and location of RPs, the SC algorithm is applied to each of the sufficiently large MUs. The proposed algorithm is shown to have a time complexity of $O((n^s)^3)$. Extensive simulations have validated the performance gain of the proposed approach over alternative approaches. The results have shown the proposed scheme to be promising by significantly reducing the number of RPs, path length, variation of RPs, and energy consumption per round. The validation of simulation results has also been shown through the statistical t-test. However, the scheme is infeasible for time-critical applications where data has to be collected from all the nodes within the desired time-bound. Therefore, we investigated the utilizing of multiple MSs for large-scale WSNs in the next chapter.

In Chapter 4, we have proposed an SSO-based scheme for two problems, one for finding an optimal number of RPs and the other for an optimal number of MS that jointly determines the traveling path of the MSs. Our scheme has the following two prominent features: firstly, it is explicitly designed for disjoint WSNs, but it is also applicable to linked networks; secondly, it is ideal for delay-tolerant applications where a shorter path length is required. Furthermore, extensive simulation conducted proved the supremacy of the scheme in terms of the number of RPs, the number of MSs, path length, and energy consumption of SNs. The results are also statistically validated through hypothesis testing and post hoc analysis.

Table 7.1: Comparison between our and conventional schemes

Metric	Conventional Schemes	Our
Network Type	Connected	Disconnected
Main goal	Energy saving	Delay mitigation
Network delay	Large	Relatively small
Communication mode	Multi-hop	Single-hop
Solution type	Heuristic/Cluster-based	Metaheuristic
Application type	Delay-lax DTNs	Delay-lax + Delay-harsh DTNs

A comparison between the proposed scheme and existing schemes is presented in Table 7.1, emphasizing that the proposed algorithm fits in delay-lax as well as delay-harsh DTNs. In such applications, a minimal path length is more important than high energy efficiency; however, reducing the delay of data collection and path length will undoubtedly increase the efficiency of such systems. In the subsequent work, we have investigated the application of data collection using MS via mobile edge computing, a state-of-the-art technology nowadays.

In Chapter 5, we have investigated the data collection problem for disjoint WSNs with mobile edge nodes and proposed a novel data collection scheme for the same by considering a realistic propagation model of the SNs. The goal is to determine a set of RPs and an optimal tour of single MS and an optimal number of MSs with their balanced trajectories. The scheme is best designed for delay-tolerant applications wherein a shorter traveling path is needed, but it is also suitable for connected networks. The scheme is composed of two algorithms; the first one is based on a convex hull for determining a set of RPs and constructing an optimal tour of a single MS, while the second one is based on the Jaya algorithm for finding an optimal number of MSs determined through subtour optimization problem. It is noteworthy that the better organization of the network via convex hull formation and an appropriate number of MSs contributes to the better performance of the proposed scheme. Through extensive simulations, we have shown that the proposed scheme can boost data collection performance compared to other existing approaches with respect to various parameters. This is also statistically validated through the ANOVA test and the plots of the confidence interval. However, the presence of obstacles in the WSNs may severely affect the performance of the proposed scheme. Therefore, the introduction of obstacles in WSN as part of IoT is next researched.

In Chapter 6, we have developed a novel scheme called OAsPP to address the path planning problem by considering a combination of a greedy approach solution to the set cover problem, which runs in polynomial time, Bezier curve, and a modified BUG2 algorithm. The objective is to plan a path for an MS in such a way that the path length is minimized and also it is smooth; the energy consumption and data collection latency are also minimized. We have shown that OAsPP is able to achieve performance gain over other relevant competing approaches with respect to various performance metrics through extensive experiments. In particular, OAsPP has been shown to achieve 10.48 % and 5.68 % improvement in terms of number of RPs and 10.12 % and 6.22 % in terms of path length over the competing approaches HOAA and OACE, respectively. It has also been shown that there has been an improvement of 16.23 % and 9.39 % in terms of total energy consumption and 6.29 % and 2.6 % with respect to network lifetime.

Herein, the different aspects of data collection are considered which addresses the questions that arise during path planning of a MS(s) in one-hop communication keeping in mind different goals as explained earlier. Further, WSNs are also an integral part of cloud edge-based IoT architecture. Thus, the above issues are also to be handled . So, various heuristic and meta-heuristic algorithms have been used owing to their individual advantages to tackle the problems. However, several issues still remain which is discussed in the following section.

7.2 Future Research Directions

The works presented in this thesis have contributed significantly towards data collection problems in WSN and then their application to mobile edge computing and IoT. Despite mitigating and solving the existing issues, several exciting issues can be further explored as follows.

- Considering the maximum number of relevant parameters simultaneously such as energy consumption, latency, and network lifetime while finding the optimal number of MSs and their paths with and without obstacles in WSNs.

- The exploration of limited mobility of SNs to improve data collection efficiency needs to be explored. Note that although this would add to the computational overhead of the data collection scheme; however, the performance of the scheme is bound to improve.

- The presence of obstacles by exploiting the restricted mobility of the SNs during data collection. Note that ostacles have irregular perimeters in real-life applications, so exploring limited mobility of SNs integrated with MS(s) can tremendously improved the performance of the scheme by contributing to the efficiency of the MS trajectory.

- The data collection in the presence of obstacles using multiple MSs is quite possible to explore. In large WSNs with obstacles single MS will lead to significant delay which may not be acceptable in some cases. Thus, employing multiple MSs will be characterized by reduced delay in data delivery to the BS.

- The use of multiple MSs in both static and dynamic obstacle embedded WSNs. Static WSNs is fairly simple to manage even in the presence of MS(s). However, in a dynamic WSN the complexity to manage the network in high as different entities involved may be mobile. For example, some SNs may be mobile, the obstacle may move, or both may have some mobility. This may be the inherent nature of the entities in vehicular-networks. It would be interesting to explore the management of the parameters involved.

- The exploration of machine learning (ML) techniques as a possible solution to the path planning problem is another potential research that can be looked into. ML techniques are known to handle several parameters simultaneously through various efficient models and provide good enough solutions, so WSN schemes can be benefited from them.

- For simplicity in some of our proposed schemes, when using multiple MSs, we have assumed that the MS and SNs remains fault-free, which may be not be be the case in reality. The fault management, i.e., fault detection and fault correction of MS(s) and SN(s) while data collection is another potential research paradigm.

- In modern and smart agriculture solutions, moisture level detection, temperature monitoring etc., WSNs coupled with ML techniques can help to design efficient data collections schemes by controlling various parameters involved.

Bibliography

[1] Francisco Javier Ferrández-Pastor, Juan Manuel García-Chamizo, Mario Nieto-Hidalgo, and José Mora-Martínez. Precision agriculture design method using a distributed computing architecture on internet of things context. *Sensors*, 18(6):1731, 2018.

[2] Sudip Misra and Amit Samanta. Traffic-aware efficient mapping of wireless body area networks to health cloud service providers in critical emergency situations. *IEEE Transactions on Mobile Computing*, 17(12):2968–2981, 2018.

[3] Kartik Trivedi and Ashish Kumar Srivastava. An energy efficient framework for detection and monitoring of forest fire using mobile agent in wireless sensor networks. In *2014 IEEE International Conference on Computational Intelligence and Computing Research*, pages 1–4. IEEE, 2014.

[4] Fariha Tasmin Jaigirdar and Mohammad Mahfuzul Islam. A new cost-effective approach for battlefield surveillance in wireless sensor networks. In *2016 International Conference on Networking Systems and Security (NSysS)*, pages 1–6. IEEE, 2016.

[5] Ismail Butun, Salvatore D Morgera, and Ravi Sankar. A survey of intrusion detection systems in wireless sensor networks. *IEEE communications surveys & tutorials*, 16(1):266–282, 2013.

[6] Halil Yetgin, Kent Tsz Kan Cheung, Mohammed El-Hajjar, and Lajos Hanzo Hanzo. A survey of network lifetime maximization techniques in wireless sensor networks. *IEEE Communications Surveys & Tutorials*, 19(2):828–854, 2017.

[7] Jamal N Al-Karaki and Ahmed E Kamal. Routing techniques in wireless sensor networks: a survey. *IEEE wireless communications*, 11(6):6–28, 2004.

[8] Xiaolong Lan, Yongmin Zhang, Lin Cai, and Qingchun Chen. Adaptive transmission design for rechargeable wireless sensor network with a mobile sink. *IEEE Internet of Things Journal*, 7(9):9011–9025, 2020.

[9] Yu Gu, Fuji Ren, Yusheng Ji, and Jie Li. The evolution of sink mobility management in wireless sensor networks: A survey. *IEEE Communications Surveys & Tutorials*, 18(1):507–524, 2015.

[10] Natalie Temene, Charalampos Sergiou, Chryssis Georgiou, and Vasos Vassiliou. A survey on mobility in wireless sensor networks. *Ad Hoc Networks*, 125:102726, 2022.

[11] Linh Nguyen and Hoc T Nguyen. Mobility based network lifetime in wireless sensor networks: A review. *Computer Networks*, 174:107236, 2020.

[12] Aiping Pang, Fan Chao, Hongbo Zhou, and Jing Zhang. The method of data collection based on multiple mobile nodes for wireless sensor network. *IEEE Access*, 8:14704–14713, 2020.

[13] Nikolaos A Pantazis, Stefanos A Nikolidakis, and Dimitrios D Vergados. Energy-efficient routing protocols in wireless sensor networks: A survey. *IEEE Communications surveys & tutorials*, 15(2):551–591, 2012.

[14] Lina Xu, Rem Collier, and Gregory MP O'Hare. A survey of clustering techniques in wsns and consideration of the challenges of applying such to 5g iot scenarios. *IEEE Internet of Things Journal*, 4(5):1229–1249, 2017.

[15] Amin Shahraki, Amir Taherkordi, Øystein Haugen, and Frank Eliassen. A survey and future directions on clustering: From wsns to iot and modern networking paradigms. *IEEE Transactions on Network and Service Management*, 18(2):2242–2274, 2020.

[16] Ruilong Deng, Shibo He, and Jiming Chen. An online algorithm for data collection by multiple sinks in wireless-sensor networks. *IEEE Transactions on Control of Network Systems*, 5(1):93–104, 2016.

[17] Sriram Chellappan, Wenjun Gu, Xiaole Bai, Dong Xuan, Bin Ma, and Kaizhong Zhang. Deploying wireless sensor networks under limited mobility constraints. *IEEE Transactions on Mobile Computing*, 6(10):1142–1157, 2007.

[18] You-Chiun Wang and Kuan-Chung Chen. Efficient path planning for a mobile sink to reliably gather data from sensors with diverse sensing rates and limited buffers. *IEEE Transactions on Mobile Computing*, 18(7):1527–1540, 2018.

[19] Qian Dong and Waltenegus Dargie. A survey on mobility and mobility-aware mac protocols in wireless sensor networks. *IEEE communications surveys & tutorials*, 15(1):88–100, 2012.

[20] Xiaojiang Ren, Weifa Liang, and Wenzheng Xu. Data collection maximization in renewable sensor networks via time-slot scheduling. *IEEE Transactions on Computers*, 64(7):1870–1883, 2014.

[21] Xuxun Liu, Tie Qiu, Bin Dai, Lei Yang, Anfeng Liu, and Jiangtao Wang. Swarm-intelligence-based rendezvous selection via edge computing for mobile sensor networks. *IEEE Internet of Things Journal*, 7(10):9471–9480, 2020.

[22] Hamidreza Salarian, Kwan-Wu Chin, and Fazel Naghdy. An energy-efficient mobile-sink path selection strategy for wireless sensor networks. *IEEE Transactions on vehicular technology*, 63(5):2407–2419, 2013.

[23] Jin Wang, Yue Yin, Jianwei Zhang, Sungyoung Lee, and R Simon Sherratt. Mobility based energy efficient and multi-sink algorithms for consumer home networks. *IEEE Transactions on Consumer Electronics*, 59(1):77–84, 2013.

[24] Hang Li and Andrey V Savkin. Wireless sensor network based navigation of micro flying robots in the industrial internet of things. *IEEE Transactions on industrial informatics*, 14(8):3524–3533, 2018.

[25] Dang Van Huynh, Tan Do-Duy, Long D Nguyen, Minh-Tuan Le, Nguyen-Son Vo, and Trung Q Duong. Real-time optimized path planning and energy consumption for data collection in unmanned ariel vehicles-aided intelligent wireless sensing. *IEEE Transactions on Industrial Informatics*, 18(4):2753–2761, 2021.

[26] Jialu Hao, Jian Liu, Wei Wu, Fengyi Tang, and Ming Xian. Secure and fine-grained self-controlled outsourced data deletion in cloud-based iot. *IEEE Internet of Things Journal*, 7(2):1140–1153, 2019.

[27] Jinke Ren, Guanding Yu, Yinghui He, and Geoffrey Ye Li. Collaborative cloud and edge computing for latency minimization. *IEEE Transactions on Vehicular Technology*, 68(5):5031–5044, 2019.

[28] Waleed Ejaz, Muhammad Naeem, and Sherali Zeadally. On-demand sensing and wireless power transfer for self-sustainable industrial internet of things networks. *IEEE Transactions on Industrial Informatics*, 17(10):7075–7084, 2020.

[29] Seung-Woo Ko, Kaifeng Han, and Kaibin Huang. Wireless networks for mobile edge computing: Spatial modeling and latency analysis. *IEEE Transactions on Wireless Communications*, 17(8):5225–5240, 2018.

[30] Nimisha Ghosh and Indrajit Banerjee. An energy-efficient path determination strategy for mobile data collectors in wireless sensor network. *Computers & Electrical Engineering*, 48:417–435, 2015.

[31] Jau-Yang Chang and Ting-Huan Shen. An efficient tree-based power saving scheme for wireless sensor networks with mobile sink. *IEEE Sensors Journal*, 16(20):7545–7557, 2016.

[32] Chien-Fu Cheng and Chao-Fu Yu. Data gathering in wireless sensor networks: a combine–tsp–reduce approach. *IEEE Transactions on Vehicular Technology*, 65(4):2309–2324, 2015.

[33] Muralitharan Krishnan, Sangwoon Yun, and Yoon Mo Jung. Enhanced clustering and aco-based multiple mobile sinks for efficiency improvement of wireless sensor networks. *Computer Networks*, 160:33–40, 2019.

[34] Wei Wang, Haoshan Shi, Dajun Wu, Pengyu Huang, Baojian Gao, Fuping Wu, Dan Xu, and Xiaojiang Chen. Vd-pso: An efficient mobile sink routing algorithm in wireless sensor networks. *Peer-to-Peer Networking and Applications*, 10(3):537–546, 2017.

[35] Charalampos Konstantopoulos, Grammati Pantziou, Damianos Gavalas, Aristides Mpitziopoulos, and Basilis Mamalis. A rendezvous-based approach enabling energy-efficient sensory data collection with mobile sinks. *IEEE Transactions on parallel and distributed systems*, 23(5):809–817, 2011.

[36] Ming Ma, Yuanyuan Yang, and Miao Zhao. Tour planning for mobile data-gathering mechanisms in wireless sensor networks. *IEEE transactions on vehicular technology*, 62(4):1472–1483, 2012.

[37] Bongsue Suh and Stevan Berber. Rendezvous points and routing path-selection strategies for wireless sensor networks with mobile sink. *Electronics Letters*, 52(2):167–169, 2015.

[38] ZhangBing Zhou, Chu Du, Lei Shu, Gerhard Hancke, Jianwei Niu, and Huansheng Ning. An energy-balanced heuristic for mobile sink scheduling in hybrid wsns. *IEEE Transactions on Industrial Informatics*, 12(1):28–40, 2015.

[39] Guoliang Xing, Tian Wang, Zhihui Xie, and Weijia Jia. Rendezvous planning in wireless sensor networks with mobile elements. *IEEE transactions on mobile computing*, 7(12):1430–1443, 2008.

[40] Xuxun Liu, Tian Wang, Weijia Jia, Anfeng Liu, and Kaikai Chi. Quick convex hull-based rendezvous planning for delay-harsh mobile data gathering in disjoint sensor networks. *IEEE Transactions on Systems, Man, And Cybernetics: Systems*, 51(6):3844–3854, 2019.

[41] Xuxun Liu, Tie Qiu, and Tian Wang. Load-balanced data dissemination for wireless sensor networks: A nature-inspired approach. *IEEE Internet of Things Journal*, 6(6):9256–9265, 2019.

[42] Xuxun Liu, Tie Qiu, Xiaobo Zhou, Tian Wang, Lei Yang, and Victor Chang. Latency-aware path planning for disconnected sensor networks with mobile sinks. *IEEE Transactions on Industrial Informatics*, 16(1):350–361, 2019.

[43] Xuxun Liu, Anfeng Liu, Tie Qiu, Bin Dai, Tian Wang, and Lei Yang. Restoring connectivity of damaged sensor networks for long-term survival in hostile environments. *IEEE Internet of Things Journal*, 7(2):1205–1215, 2019.

[44] Praveen Kumar, Tarachand Amgoth, and Chandra Sekhara Rao Annavarapu. Aco-based mobile sink path determination for wireless sensor networks under non-uniform data constraints. *Applied Soft Computing*, 69:528–540, 2018.

[45] Hailong Huang, Chao Huang, and Dazhong Ma. The cluster based compressive data collection for wireless sensor networks with a mobile sink. *AEU-International Journal of Electronics and Communications*, 108:206–214, 2019.

[46] Can Tunca, Sinan Isik, Mehmet Yunus Donmez, and Cem Ersoy. Ring routing: An energy-efficient routing protocol for wireless sensor networks with a mobile sink. *IEEE Transactions on Mobile Computing*, 14(9):1947–1960, 2014.

[47] YoungSang Yun and Ye Xia. Maximizing the lifetime of wireless sensor networks with mobile sink in delay-tolerant applications. *IEEE Transactions on mobile computing*, 9(9):1308–1318, 2010.

[48] Hee-Sook Mo, Euisin Lee, Soochang Park, and Sang-Ha Kim. Virtual line-based data dissemination for mobile sink groups in wireless sensor networks. *IEEE Communications Letters*, 17(9):1864–1867, 2013.

[49] Shubhra Jain, KK Pattanaik, Rahul Kumar Verma, Sourabh Bharti, and Anupam Shukla. Delay-aware green routing for mobile-sink-based wireless sensor networks. *IEEE Internet of Things Journal*, 8(6):4882–4892, 2020.

[50] Zhang Lin, Huan-Chao Keh, Ruikun Wu, and Diptendu Sinha Roy. Joint data collection and fusion using mobile sink in heterogeneous wireless sensor networks. *IEEE Sensors Journal*, 21(2):2364–2376, 2020.

[51] Chu-Fu Wang, Jau-Der Shih, Bo-Han Pan, and Tin-Yu Wu. A network lifetime enhancement method for sink relocation and its analysis in wireless sensor networks. *IEEE sensors journal*, 14(6):1932–1943, 2014.

[52] Ye Miao, Zhili Sun, Ning Wang, Yue Cao, and Haitham Cruickshank. Time efficient data collection with mobile sink and vmimo technique in wireless sensor networks. *IEEE Systems Journal*, 12(1):639–647, 2016.

[53] Arun A Somasundara, Aditya Ramamoorthy, and Mani B Srivastava. Mobile element scheduling with dynamic deadlines. *IEEE transactions on Mobile Computing*, 6(4):395–410, 2007.

[54] Ryo Sugihara and Rajesh K Gupta. Optimal speed control of mobile node for data collection in sensor networks. *IEEE Transactions on Mobile Computing*, 9(1):127–139, 2009.

[55] Bo Yuan, Maria Orlowska, and Shazia Sadiq. On the optimal robot routing problem in wireless sensor networks. *IEEE transactions on knowledge and data engineering*, 19(9):1252–1261, 2007.

[56] Theodore S Rappaport et al. *Wireless communications: principles and practice*, volume 2. prentice hall PTR New Jersey, 1996.

[57] Can Tunca, Sinan Isik, M Yunus Donmez, and Cem Ersoy. Distributed mobile sink routing for wireless sensor networks: A survey. *IEEE communications surveys & tutorials*, 16(2):877–897, 2013.

[58] Soochang Park, Euisin Lee, Min-Sook Jin, and Sang-Ha Kim. Novel strategy for data dissemination to mobile sink groups in wireless sensor networks. *IEEE Communications Letters*, 14(3):202–204, 2010.

[59] Robert Tarjan. Depth-first search and linear graph algorithms. *SIAM journal on computing*, 1(2):146–160, 1972.

[60] Ulrike Von Luxburg. A tutorial on spectral clustering. *Statistics and computing*, 17(4):395–416, 2007.

[61] Jianbo Shi and Jitendra Malik. Normalized cuts and image segmentation. *IEEE Transactions on pattern analysis and machine intelligence*, 22(8):888–905, 2000.

[62] Uporabo Statisti and N Tehnike. Improving the process capability of a turning operation by the application of statistical techniques. *Materiali in tehnologije*, 43(1):55–59, 2009.

[63] Oveis Abedinia, Nima Amjady, and Ali Ghasemi. A new metaheuristic algorithm based on shark smell optimization. *Complexity*, 21(5):97–116, 2016.

[64] Klaus Hinkelmann, Oscar Kempthorne, Klaus Hinkelmann, and Oscar Kempthorne. Design and analysis of experiments: Advanced experimental design, volume 2. 2005.

[65] R Rao. Jaya: A simple and new optimization algorithm for solving constrained and unconstrained optimization problems. *International Journal of Industrial Engineering Computations*, 7(1):19–34, 2016.

[66] Vladimir J Lumelsky and Alexander A Stepanov. Path-planning strategies for a point mobile automaton moving amidst unknown obstacles of arbitrary shape. *Algorithmica*, 2(1):403–430, 1987.

[67] Jin Wang, Yiquan Cao, Bin Li, Hye-jin Kim, and Sungyoung Lee. Particle swarm optimization based clustering algorithm with mobile sink for wsns. *Future Generation Computer Systems*, 76:452–457, 2017.

[68] Jun Tao, Liang He, Yanyan Zhuang, Jianping Pan, and Maryam Ahmadi. Sweeping and active skipping in wireless sensor networks with mobile elements. In *2012 IEEE Global Communications Conference (GLOBECOM)*, pages 106–111. IEEE, 2012.

[69] Khaled Almi'ani, Anastasios Viglas, and Lavy Libman. Energy-efficient data gathering with tour length-constrained mobile elements in wireless sensor networks. In *IEEE local computer network conference*, pages 582–589. IEEE, 2010.

[70] Russell Eberhart and James Kennedy. Particle swarm optimization. In *Proceedings of the IEEE international conference on neural networks*, volume 4, pages 1942–1948. Citeseer, 1995.

[71] Wei Hong Lim and Nor Ashidi Mat Isa. Particle swarm optimization with increasing topology connectivity. *Engineering Applications of Artificial Intelligence*, 27:80–102, 2014.

[72] Marco Dorigo, Vittorio Maniezzo, and Alberto Colorni. Ant system: optimization by a colony of cooperating agents. *IEEE Transactions on Systems, Man, and Cybernetics, Part B (Cybernetics)*, 26(1):29–41, 1996.

[73] John Henry Holland et al. *Adaptation in natural and artificial systems: an introductory analysis with applications to biology, control, and artificial intelligence*. MIT press, 1992.

[74] Emo Welzl. Smallest enclosing disks (balls and ellipsoids). In *New results and new trends in computer science*, pages 359–370. Springer, 1991.

[75] Miao Zhao, Yuanyuan Yang, and Cong Wang. Mobile data gathering with load balanced clustering and dual data uploading in wireless sensor networks. *IEEE Transactions on Mobile Computing*, 14(4):770–785, 2014.

[76] Saamaja Vupputuri, Kiran K Rachuri, and C Siva Ram Murthy. Using mobile data collectors to improve network lifetime of wireless sensor networks with reliability constraints. *Journal of Parallel and Distributed Computing*, 70(7):767–778, 2010.

[77] Hari Prabhat Gupta, SV Rao, Amit Kumar Yadav, and Tanima Dutta. Geographic routing in clustered wireless sensor networks among obstacles. *IEEE sensors Journal*, 15(5):2984–2992, 2014.

[78] Ramesh Kumar, Tarachand Amgoth, and Debjit Das. Obstacle-aware connectivity establishment in wireless sensor networks. *IEEE Sensors Journal*, 21(4):5543–5552, 2020.

[79] Zhenqiang Mi, Yang Yang, and James Yifei Yang. Restoring connectivity of mobile robotic sensor networks while avoiding obstacles. *IEEE Sensors Journal*, 15(8):4640–4650, 2015.

[80] Jiang Shao, Delin Luo, Yang Xu, and Haibin Duan. Cooperative path planning for multiple robots with motion constraints in obstacle-strewn environment. *IEEE Access*, 7:132286–132301, 2019.

[81] Gregor Klančar, Marija Seder, Sašo Blažič, Igor Škrjanc, and Ivan Petrović. Drivable path planning using hybrid search algorithm based on e* and bernstein-bezier motion primitives. *IEEE Transactions on Systems, Man, and Cybernetics: Systems*, 51(8):4868–4882, 2019.

[82] Satish Sundar and Zvi Shiller. Optimal obstacle avoidance based on the hamilton-jacobi-bellman equation. *IEEE transactions on robotics and automation*, 13(2):305–310, 1997.

[83] Guangqian Xie and Feng Pan. Cluster-based routing for the mobile sink in wireless sensor networks with obstacles. *IEEE Access*, 4:2019–2028, 2016.

[84] Tomas Berglund, Andrej Brodnik, Håkan Jonsson, Mats Staffanson, and Inge Soderkvist. Planning smooth and obstacle-avoiding b-spline paths for autonomous mining vehicles. *IEEE Transactions on Automation Science and Engineering*, 7(1):167–172, 2009.

[85] S JL Diaz de Leon and A JH Sossa. Automatic path planning for a mobile robot among obstacles of arbitrary shape. *IEEE Transactions on Systems, Man, and cybernetics. Part B, Cybernetics: a Publication of the IEEE Systems, Man, and Cybernetics Society*, 28(3):467–472, 1998.

[86] Joo Young Hwang, Jun Song Kim, Sang Seok Lim, and Kyu Ho Park. A fast path planning by path graph optimization. *IEEE Transactions on systems, man, and cybernetics-part a: systems and humans*, 33(1):121–129, 2003.

[87] Hamid Mahboubi, Walid Masoudimansour, Amir G Aghdam, and Kamran Sayrafian-Pour. Maximum lifetime strategy for target monitoring with controlled node mobility in sensor networks with obstacles. *IEEE transactions on automatic control*, 61(11):3493–3508, 2016.

[88] Guang Tan, Stephen A Jarvis, and Anne-Marie Kermarrec. Connectivity-guaranteed and obstacle-adaptive deployment schemes for mobile sensor networks. *IEEE Transactions on Mobile Computing*, 8(6):836–848, 2009.

[89] Samad Najjar-Ghabel, Leili Farzinvash, and Seyed Naser Razavi. Mobile sink-based data gathering in wireless sensor networks with obstacles using artificial intelligence algorithms. *Ad Hoc Networks*, 106:102243, 2020.

[90] Li Liu, Guangjie Han, Hao Wang, and Jiafu Wan. Obstacle-avoidance minimal exposure path for heterogeneous wireless sensor networks. *Ad Hoc Networks*, 55:50–61, 2017.

[91] Jian Wang, Zhengxing Wu, Min Tan, and Junzhi Yu. 3-d path planning with multiple motions for a gliding robotic dolphin. *IEEE Transactions on Systems, Man, and Cybernetics: Systems*, 51(5):2904–2915, 2019.

[92] Yanli Chen, Guiqiang Bai, Yin Zhan, Xinyu Hu, and Jun Liu. Path planning and obstacle avoiding of the usv based on improved aco-apf hybrid algorithm with adaptive early-warning. *IEEE Access*, 9:40728–40742, 2021.

[93] Marco Dorigo and Thomas Stützle. Ant colony optimization: overview and recent advances. *Handbook of metaheuristics*, pages 311–351, 2019.

[94] Xiaolin He, Xiuwen Fu, and Yongsheng Yang. Energy-efficient trajectory planning algorithm based on multi-objective pso for the mobile sink in wireless sensor networks. *IEEE Access*, 7:176204–176217, 2019.

[95] Yinggao Yue, Li Cao, Bo Hang, and Zhongqiang Luo. A swarm intelligence algorithm for routing recovery strategy in wireless sensor networks with mobile sink. *IEEE Access*, 6:67434–67445, 2018.

[96] Xiuwen Fu and Xiaolin He. Energy-balanced data collection with path-constrained mobile sink in wireless sensor networks. *AEU-International Journal of Electronics and Communications*, 127:153504, 2020.

[97] Anis Jari and Avid Avokh. Pso-based sink placement and load-balanced anycast routing in multi-sink wsns considering compressive sensing theory. *Engineering Applications of Artificial Intelligence*, 100:104164, 2021.

[98] Manish Kumar Singh, Syed Intekhab Amin, and Amit Choudhary. Genetic algorithm based sink mobility for energy efficient data routing in wireless sensor networks. *AEU-International Journal of Electronics and Communications*, 131:153605, 2021.

[99] Xuxun Liu, Peihang Lin, Tang Liu, Tian Wang, Anfeng Liu, and Wenzheng Xu. Objective-variable tour planning for mobile data collection in partitioned sensor networks. *IEEE Transactions on Mobile Computing*, 21(1):239–251, 2020.

[100] David S Johnson, Gregory Gutin, Lyle A McGeoch, Anders Yeo, Weixiong Zhang, and Alexei Zverovitch. Experimental analysis of heuristics for the atsp. In *The traveling salesman problem and its variations*, pages 445–487. Springer, 2007.

[101] Liang He, Jianping Pan, and Jingdong Xu. Reducing data collection latency in wireless sensor networks with mobile elements. In *2011 IEEE Conference on Computer Communications Workshops (INFOCOM WKSHPS)*, pages 572–577. IEEE, 2011.

[102] B Heinzelman, Wendi, P Chandrakasan, Anantha, and Hari Balakrishnan. An application-specific protocol architecture for wireless microsensor networks. *IEEE Transactions on wireless communications*, 1(4):660–670, 2002.

[103] Ibrahiem MM El Emary and S Ramakrishnan. *Wireless sensor networks: from theory to applications*. CRC press, 2019.

[104] Peng Xiao, Wenyu Qu, Heng Qi, Zhiyang Li, and Yujie Xu. The sdn controller placement problem for wan. In *2014 IEEE/CIC International Conference on Communications in China (ICCC)*, pages 220–224. IEEE, 2014.

[105] Donghui Yan, Ling Huang, and Michael I Jordan. Fast approximate spectral clustering. In *Proceedings of the 15th ACM SIGKDD international conference on Knowledge discovery and data mining*, pages 907–916, 2009.

[106] Bong Sun Kim, Sang Gyu Park, Young Kwan You, and Soo Il Jung. Probability & statistics for engineers & scientists, 2011.

[107] Meenalochani Manickam and Sudha Selvaraj. Range-based localisation of a wireless sensor network using jaya algorithm. *IET Science, Measurement & Technology*, 13(5):678–684, 2019.

[108] Chao Huang, Long Wang, Ryan Shun-Cheung Yeung, Zijun Zhang, Henry Shu-Hung Chung, and Alain Bensoussan. A prediction model-guided jaya algorithm for the pv system maximum power point tracking. *IEEE Transactions on sustainable energy*, 9(1):45–55, 2017.

[109] Prateek Raj Gautam, Sunil Kumar, Akshay Verma, Tarique Rashid, and Arvind Kumar. Energy-efficient localization of sensor nodes in wsns using beacons from rotating directional antenna. *IEEE Transactions on Industrial Informatics*, 15(11):5827–5836, 2019.

[110] Mou Wu, Naixue Xiong, and Liansheng Tan. Adaptive range-based target localization using diffusion gauss–newton method in industrial environments. *IEEE Transactions on Industrial Informatics*, 15(11):5919–5930, 2019.

[111] Franco P Preparata and Michael I Shamos. *Computational geometry: an introduction*. Springer Science & Business Media, 2012.

[112] William C Jakes and Donald C Cox. *Microwave mobile communications*. Wiley-IEEE Press, 1994.

[113] Thomas H Cormen, Charles E Leiserson, Ronald L Rivest, and Clifford Stein. *Introduction to algorithms*. MIT press, 2009.

[114] Ray Galvin. Energy consumption effects of speed and acceleration in electric vehicles: Laboratory case studies and implications for drivers and policymakers. *Transportation Research Part D: Transport and Environment*, 53:234–248, 2017.

[115] Keith E Muller and Bethel A Fetterman. *Regression and ANOVA: an integrated approach using SAS software*. John Wiley & Sons, Inc., 2003.

[116] Yutaka J Kanayama and Bruce I Hartman. Smooth local-path planning for autonomous vehicles1. *The International Journal of Robotics Research*, 16(3):263–284, 1997.

[117] Hongluo Li, Yutao Luo, and Jie Wu. Collision-free path planning for intelligent vehicles based on bézier curve. *IEEE Access*, 7:123334–123340, 2019.

[118] Jon Kleinberg and Eva Tardos. *Algorithm design*. Pearson Education India, 2006.

CPSIA information can be obtained
at www.ICGtesting.com
Printed in the USA
LVHW020013280423
745413LV00009B/223

9 781805 258797